THE GREAT MELT

THE GREAT MELT

ACCOUNTS FROM THE FRONTLINE OF CLIMATE CHANGE

ALISTER DOYLE

FLINT

For Siv, Emma
and Matias

First published 2021

FLINT is an imprint of The History Press
97 St George's Place, Cheltenham,
Gloucestershire, GL50 3QB
www.flintbooks.co.uk

British Library Cataloguing in Publication Data.
A catalogue record for this book is available from the British Library.

ISBN 978 0 7509 9784 3

Typesetting and origination by The History Press
Printed and bound in Great Britain by TJ Books Limited, Padstow, Cornwall.

Maps © Geethik Technologies, India.

Trees for Life

CONTENTS

FOREWORD

BY CHRISTIANA FIGUERES

IN 1962, AN AMERICAN OIL company proudly told readers of *Life Magazine* that it "supplies enough energy to melt 7 million tons of glacier" every day.

"This giant glacier has remained unmelted for centuries," Humble Oil said in an advertisement by a photo of a frozen river of ice in Alaska. It described how the firm "provides energy in many forms – to help heat our homes, power our transportation and to furnish industry with a great variety of versatile chemicals".

How times have changed since that two-page boast about the power of human ingenuity to alter the natural world was published, in an edition of *Life* that featured astronaut John Glenn on the cover.

We now know that greenhouse gases released by burning fossil fuels in factories, cars and power plants are the main driver of an accelerating rise in temperatures that scientists estimate is thawing a staggering 2 billion tonnes of ice on land every day, from Greenland to Antarctica.

That meltwater is cascading into the oceans and raising sea levels – in the worst case by 7 metres by 2300 that would swamp coasts from Bangladesh to Miami, and entire low-lying nations in the

Pacific and Indian Oceans, forcing tens if not hundreds of millions of people to leave their homes and migrate to what will most likely be unwelcoming regions.

In *The Great Melt*, Alister Doyle gives voice to people living on the frontlines of climate change, as it is happening right now, in a stark reminder that we are at a crossroads in the climate crisis. The Sixth Assessment Report of the Intergovernmental Panel on Climate Change (IPCC) leaves no doubt that this is the decisive decade for humankind in which we need to halve emissions by 2030 from 2010 levels to avoid the worst impacts of warming.

Alister visits a Fijian family who have moved inland – not once, but twice – to escape rising seas. In Iceland, a poet is struggling with the loss of glaciers that she thought were “emblems of eternity”. In Panama, a man is alarmed by seeing sardines swimming in his house during a storm surge. In Florida, a teenage activist is trying to use the courts to avert what could be a metre of sea level rise during his lifetime. Alister also flies with British scientists to land on a part of Antarctica – the Wilkins Ice Shelf – that has since shattered and vanished in the Southern Ocean. These personal stories are juxtaposed with the latest science to create a powerful base of evidence for urgent action to avert further suffering.

I was Executive Secretary of the UN Framework Convention on Climate Change from 2010 to 2016, during most of that time we worked toward the Paris Agreement after the debacle of the Copenhagen climate summit in 2009. In *The Future We Choose*, the 2020 book I wrote with my friend and colleague Tom Rivett-Carnac, we warn of the apocalyptic risks of sea level rise if we fail to achieve the Paris goals of limiting rising temperatures. If the West Antarctic Ice Sheet collapses, for instance, we imagine coastal mega-cities as “ghostly Atlantises dotting the coasts of each continent, their skyscrapers jutting out of the water, their people evacuated or dead”. But we are not condemned to this devastation. As *The Great*

Melt also underscores, it is not too late: we can not only survive the climate crisis, we can thrive in the better world we intentionally, but urgently, create.

I have appreciated Alister's excellent coverage of UN negotiations on climate change for almost two decades, mostly in his former job as environment correspondent at Reuters. We have met in places from Lima to Paris, from Geneva to Cancun, often in conference halls where governments hammer out legal texts meant to safeguard the planet and the most vulnerable people on the frontlines.

There are many reasons for both outrage and optimism. Outrage at the slow pace of action and the rapid pace of destruction, optimism because many things are changing – we have the blueprint for action in the 2015 Paris Agreement and most of the world's population now lives in nations with targets for net zero emissions, often by mid-century. Governments, businesses, investors, voters and young people are becoming ever more aware of the need to act.

As we seek a greener economic future after the coronavirus pandemic, the pace of the melt of glaciers in coming decades will be one of the clearest gauges of our success or failure in fighting global warming. This book is an innovative wake-up call for action from the fragile frontlines.

Christiana Figueres
Executive Secretary of the UN Framework Convention on Climate Change, 2010–16, Co-Founder of Global Optimism
August 2021

JOURNEYS INTO THE FRONTLINES OF CLIMATE CHANGE

ON A BEACH IN EASTERN Fiji shaded by coconut palms, the chief of the local village stands dejectedly beside rotting wooden stumps and a lump of concrete jutting out of the sand. "This is my house," Simione Botu, a stocky man in his early sixties, tells me with a shrug. Living on a frontline of climate change, this is all that remains of his boyhood home by the Pacific Ocean, washed away decades ago.

In a world where sea level rise driven by global warming is melting ever more ice from Antarctica to Greenland and threatens cities from San Francisco to Shanghai, Botu and his family are among a vanishingly small group: double victims forced to rebuild their homes inland not just once, but twice.

After storm surges wrecked his childhood home near the mouth of a meandering river, Botu built a new one-storey home 50 metres inland. But makeshift sea walls failed, and the water followed, damaging the wooden foundations. In 2014 the Fijian government, with rare foresight, relocated his entire village of 150 people to new blue-painted homes, on a hillside more than 1km inland.

On the other side of the world, where the coast of Florida gets flooded when hurricanes whip up the Atlantic Ocean, Levi Draheim tells me he feels "sadness and anger" that his young sister will inherit a planet where the seas could rise by a metre this century.

He has already had to wade in water up to his knees when his home was flooded. "I want to be able to tell her that I did everything I can to fight these horrible things that we are inheriting," he said. His compassion is striking: when we talk in 2021, Draheim is aged 13 and his sister, Juniper, a nine-month-old infant. Levi Draheim is a plaintiff in a lawsuit trying to force Florida to take far more action to cut fossil fuel use. He is part of a global surge of activism by young people, largely inspired by Sweden's Greta Thunberg and her #FridaysforFuture movement.

And alarm about climate change is rising. In the starkest warning to date about climate change, top scientists and governments issued a UN report in August 2021[1] that UN Secretary-General António Guterres called "code red for humanity".[2]

The Intergovernmental Panel on Climate Change (IPCC), the world's most authoritative guide to global warming comprising both governments and scientists, concluded for the first time that the evidence is "unequivocal" – beyond a shadow of doubt – that humans are to blame for global warming that is causing fiercer heatwaves, downpours, wildfires and a thaw of ice that is driving up the oceans.

For many people alive today, sea level rise seems a distant threat, beyond the lifetimes of current policymakers. And yet many living today are already suffering the consequences of a global increase of the oceans, of about 20cm since 1900. And Juniper Draheim will be 79 on New Year's Day in 2100.

Botu and Draheim are among the people living on the frontlines of climate change who have helped me to write this book, based on visits to places including Antarctica, Peru, Panama, Fiji, the Netherlands, Britain and Sweden, backed up with calls during the pandemic. *The Great Melt* is meant to offer a snapshot of how people on the frontlines are managing the crisis of melting ice and sea level rise, with both successes and failures.

I set out trying to understand: when seas are rising worldwide, why are some people much more at risk than others? How much will seas rise this century, and beyond? Other than cutting greenhouse gas emissions, how do you protect a coastline? How do people make the agonising decision to move inland, in some cases leaving behind the graves of their ancestors? Can you go to court to blame someone? Can you be a 'climate refugee'? Can a low-lying island nation continue to exist if it disappears beneath the waves, like a modern-day Atlantis?

Along the way are tales of resilience. An elderly Panamanian man, who wants to move to the mainland from a low-lying island, laughs when he recounts his dismay at seeing sardines swimming in his home during a tropical storm sweeping the Caribbean. An Icelandic poet bemoans the accelerating loss of a glacier that she remembers as an "emblem of eternity" when she led cows past the ice to summer pastures as a girl. In Peru, I meet a mountain guide who is suing a German power company on the other side of the world. He blames the company for contributing to glacier melt high in the Andes, threatening a mudslide that could wipe away his family home. In Fiji in the South Pacific Ocean, I meet people torn between two identities after their parents and grandparents were forcibly relocated in 1945 from an island 2,000km away after it was ruined by British phosphate mining. The struggle by the displaced Banabans to secure rights in Fiji, such as land ownership and citizenship, is a rare, imperfect blueprint for future migration driven by climate change.

Apart from people living on the frontlines, I have spoken to scientists, lawyers, government officials, environmentalists, activists and artists trying to make sense of the bewildering melt. Dozens of people have generously given me their time.

Some climate sceptics say sea level rise is exaggerated: the sea is constantly shifting – the tides rise and fall, storms come and go, currents swirl. They might argue that the coastline by the white cliffs in what is now Kent in the south of England probably looks pretty much as it did when Julius Caesar invaded in 55BC, and that coasts subside because of natural geological forces and erosion, such as homes on islands in the Chesapeake Bay in the US abandoned a century ago.[3]

They could also point to cities such as Jakarta in Indonesia that are sinking as they suck water from aquifers deep below.

Yet a momentous change is under way in the oceans that cover 70 per cent of the planet's surface – they are undoubtedly warming because of our greenhouse gas emissions. The waters are stirring as glaciers from the Alps to the Andes are melting and the vast ice sheets of Greenland and Antarctica thaw.

The centuries since Caesar invaded Britain marked a rare period of stability in sea levels[4] during Earth's history and allowed many civilisations to flourish by the coasts. That stability is ending, after the Industrial Revolution in the eighteenth century ushered in the increasing use of coal and other fossil fuels.

Annual sea level rise has surged to 3.7mm a year in recent years, more than double the average rate for the twentieth century, the IPCC report says.

The amounts of ice are mind-boggling: every mm of sea level rise is equivalent to melting an ice cube with sides about 7km long – roughly 360 billion tonnes of ice. Such an ice cube would be as tall as Aconcagua, the highest mountain in the Andes. Scientists sometimes compare the impact of the Industrial Revolution to a mammoth hair dryer blowing hot air onto an ice cube.

The fate of the world's coasts now rests on a knife edge. Choices made this decade at meetings including the global climate summit in Glasgow, Scotland, in November 2021 (known as COP26), will determine whether the rise of the ocean remains within manageable limits in coming centuries, or redraws maps of the world.

But I've seen that happening already. In Antarctica, I accompanied scientists from the British Antarctic Survey to a part of the world that is no longer on the map – the Wilkins Ice Shelf broke up weeks after we landed in a tiny plane equipped with skis in a nail-biting flight in 2009.

Keep pumping ever more greenhouse gases into the atmosphere, the IPCC says, and a melt is likely to drive up sea levels by up to a metre by 2100 and by between 2 and 7 metres by 2300. Seven metres is taller than a giraffe, above the world record for the men's pole vault, or about the height of a two-storey building. That Pandora's Box future would swamp coastal areas from Bangladesh to Florida and entire island nations in the Pacific or Indian Oceans. But there is still hope. Act quickly now to limit greenhouse gas emissions, however, and seas might rise by in the best case by only about 30cm this century and less than a metre by 2300. That would be bad, but fixable.

Outside these core scenarios, however, seas could rise by almost two metres by 2100 in the unlikely case that Antarctica's ice sheet starts to disintegrate, it says. And a disturbing note about "deep uncertainties" for 2300, says: "Sea level rise greater than 15 metres cannot be ruled out with high emissions." In Washington DC, that would raise the Atlantic Ocean to the lawn of the White House. The report, by more than 200 scientists, is based on 14,000 scientific papers, and endorsed by governments.

Some are suggesting building new walls, perhaps dumping vast amounts of sand on the coastlines to protect cities, as is already happening in the Netherlands. Meanwhile, around the world, people on the frontlines face a host of legal difficulties: who will go where? Who will pay? Are women, young people and minorities getting enough say? *The Great Melt* is a stark reminder for a world struggling with Covid-19 that greenhouse gases, as invisible and as insidious as the virus, are an existential threat for coming centuries.

The idea for this book came from fifteen years working as environment correspondent, from 2004 to 2019, for Thomson Reuters. I attended dozens of climate conferences – including thirteen of the annual COPs in places from Milan to Bali, from Cancun to Nairobi. I heard low-lying island states, from the Pacific to the Indian Ocean, complain that big nations often dismissed their concerns about sea levels as affecting only a miniscule fraction of humanity. At a UN climate summit in Copenhagen in 2009, for instance, Dessima Williams of the island nation of Grenada in the Caribbean berated the Danish organisers for hanging a huge globe, several metres across, in the cavernous conference centre. She was outraged because the black and white map was stylised and omitted islands smaller than about the size of Malta, meaning entire nations had vanished into oblivion in the blank ocean. One frustrated Pacific island delegate had added dots with a pencil and written in the name of his homeland – the Solomon Islands. "We need to be on that map," Williams said.

So, I have made a point of trying to visit remote places, including some left off that Copenhagen map, to see how sea level rise is affecting lives there. Since Copenhagen, sea level rise has become a lot more urgent, with both rich and poor nations waking up to the risks and costs of losing coastlines.

"The alarm bells are deafening, and the evidence is irrefutable," Guterres said of the IPCC's findings. "This report must sound a death knell for coal and fossil fuels, before they destroy our planet. There must be no new coal plants built after 2021." He also wants governments, focused since 2020 on the coronavirus pandemic, to address climate change which he describes as "another, deep emergency".

Guterres says governments should halve greenhouse gas emissions from 2010 levels by 2030, requiring cuts of 7.6 per cent a year throughout this decade, as we shift away from the fossil fuel economy. But climate action is a perplexing rollercoaster – news is sometimes inspiring, often depressing. And bold promises for cuts in emissions usually far outstrip action.

One major cause for optimism is that almost 200 governments adopted the historic Paris Agreement in 2015 to end the fossil fuel era this century, and President Joe Biden rejoined in 2021 after former President Donald Trump pulled out. The agreement seeks to limit global warming to "well below" 2°C above pre-industrial times, while "pursuing efforts" for an even tougher goal of 1.5°C. Since that breakthrough, most of the world's population lives in nations with a goal of reaching net zero emissions by 2050 or 2060. Equally important, prices of solar and wind power have plunged, and investors and companies are shifting to greener technologies, while youth activists and groups such as Extinction Rebellion have added pressure. In July 2021, the European Union policymakers laid out their boldest climate plan yet, overhauling everything from industry to shipping in order to slash EU emissions by 55 per cent from 1990 levels by 2030. In 2020, global emissions of carbon dioxide tumbled a record 6 per cent,[5] the largest annual drop since the Second World War and close to the UN goal of 7.6 per cent every year this decade.

But the bad news is that emissions are expected to leap 5 per cent in 2021 as demand for coal, oil and gas rebounds. All the while, almost no politicians have legislated short-term goals to cut emissions in coming years to get on track for the Paris Agreement, and many corporate commitments to sustainability are hyperbolic. Meanwhile, 2020 was one of the three warmest years on record,[6] along with 2016 and 2019, with temperatures about 1.2°C above pre-industrial times. Record heatwaves baked Canada and the United States in 2021, an event climate researchers called 'virtually impossible' without human-caused climate change.[7] And deadly floods swamped parts of Europe, China and India. Elsewhere, devastating wildfires ravaged parts of Siberia, Greece and California and a heatwave dubbed "Lucifer" seared southern Italy.

Nowhere has the contrast between whether to act, dither or simply ignore climate change been starker than in the US – the world's biggest economy and number two greenhouse gas emitter behind China. Biden pledged in 2021 to halve US emissions by 2030, from 2005 levels, and said it would help the economy. "For too long we've failed to use the most important word when it comes to meeting the climate crisis: Jobs. Jobs. Jobs. For me, when I think climate change, I think jobs," he told the US Congress.

But Washington has a whiplash history of climate policies. A Chinese foreign ministry spokesperson said the US embrace of the Paris Agreement "is not the return of the King, rather it's a truant getting back to class". So, the outlook for sea levels is uncertain.

Trump, under whose leadership the US was the only nation to quit the 2015 agreement, once tweeted that a massive sea wall to protect New York from rare storms would be "a costly, foolish & environmentally unfriendly idea ... Sorry, you'll just have to get your mops & buckets ready!" By contrast in 2008, on securing the Democratic Party nomination to run for president, Barack Obama had expressed soaring hopes that "we will be able to look back and tell our children that this was the moment when the rise of the oceans began to slow, and our planet began to heal." Still, Biden's plans are a huge cause for new hope.

"This is the decisive decade," said Christiana Figueres, who was an architect of the Paris Agreement as the UN's former climate chief. Biden's plans "keep the 1.5°C goal alive for us. That is absolutely critical." Laurence Tubiana, who also designed the Paris Agreement as France's climate ambassador, welcomed Biden's goals as a "huge leap" but said much more was needed by the whole world, including new funds for developing nations.

Despite decades of promises of more action, the world's greenhouse gas emissions have risen almost every year since 1990 (the baseline set at an Earth Summit attended by US Republican President George H.W. Bush). The consequences are ever more visible, from decaying coral reefs to loss of forests and wildlife, as the planet heats up.

The big worries are the ice sheets in a warming world. Antarctica contains enough ice to drive up sea levels by 58 metres and Greenland locks up the equivalent of 7 metres. There are alarming hints that an 'irreversible' melt may already be under way. But to truly understand the context of this melt, we have to go back in time.

SEA LEVEL RISE: AN OLD STORY

For much of history, changes in sea levels have been an almost constant feature of lives along the coasts – but the last 2,000 years have been exceptionally stable.

In the biblical story of Noah's Ark – echoing the flood myth in the *Epic of Gilgamesh* in ancient Mesopotamia – God sends rains that swamp most of the planet. Noah builds a vast ark to save his family and pairs of each type of creature to ensure a new start for life on Earth.

Sea levels have risen about 120 metres in the past 20,000 years, as the Ice Age loosened its grip. A vast pile of sand dredged up from the North Sea to reinforce a beach in the Netherlands, for instance, contains fossils of mammoths and deer – relics of a time when the region was dry land and Britain was connected to continental Europe.

Seas are rising almost everywhere, but I also visit parts of Scandinavia where they are falling relative to the land. Along the Baltic Sea, the coast is rebounding since the end of the Ice Age lifted a vast frozen weight off the land, like a huge foam mattress that takes a while to reshape after you get up. *The Great Melt* uncovers a detective story dating back to the eighteenth century about how scientists solved a mystery of the falling level of the Baltic Sea, helped by Swedish 'seal rocks' and an ancient oak tree, still standing near Stockholm.

In this book, I have tried to look around the globe to highlight the harrowing choices faced by those on the frontline of climate change. It's involved a lot of travel, by air, buses, trucks, trains, cars and on foot: I have bought offsets for the carbon emissions and hope this book justifies the pollution.

Beyond the suffering of people on the front lines who will be forced to move, Anders Levermann, a professor at the Potsdam Institute for Climate Impact Research and Columbia University, said that several metres of sea level rise would also imperil many of humanity's greatest cultural achievements, from Venice to Sydney, from Beijing to London. "So many cities will be seriously threatened ... It's a threat to our cultural heritage," he said.

Asked about the major events of the twentieth century in opinion polls, people in Western nations often mention events such as the Second World War, the rise and fall of the Soviet Union, Neil Armstrong and Buzz Aldrin's landing on the Moon, votes for women, the Holocaust, or the atom bomb. Ask the same question in the year 2300 and, in the worst case, people may well remember our generation for the climate crisis and how we redrew maps of the planet as the seas rose. We can still avert the worst legacy of sea level rise. Future generations would not forgive us a nation-swamping 7 metres.

1

IN ANTARCTICA, THE GIANT STIRS

The ice was here, the ice was there,
The ice was all around:
It cracked and growled, and roared and howled,
Like noises in a swound!

The Rime of the Ancient Mariner, Samuel Taylor Coleridge (1834)

Around the fringes of Antarctica, by far the biggest store of ice on the planet, glaciers are showing signs of an 'irreversible' melt that could swamp the world's coasts. The focus of research is on a huge 'doomsday glacier', the Thwaites, after the break-up of ice shelves, including the Wilkins, to the north. Scientists, and other staff including cooks, pilots and electricians, are devoting their careers to understanding a continent that could be a frozen 'time bomb' for sea level rise.

A BUZZER SOUNDS IN THE tiny red plane as it swoops down to land on the Wilkins Ice Shelf off Antarctica, unnerving the first – and only – visitors to a part of the world that has since vanished off the map in a chaos of icebergs.

Seals lolling on the sea ice look up as we pass above them, mystified by the whirr of what is almost certainly the only thing they've ever seen flying apart from birds. 'Remote' hardly captures how far this white slab of ice is from civilisation – the nearest town is Punta Arenas in Chile more than 2,000km to the north, beyond the stormy Southern Ocean.

As the alarm sounds, scientists from the British Antarctic Survey and we two reporters aboard brace for the worst: I grip the metal frames of the bare-bones seat of the Twin Otter plane, equipped with skis as well as wheels, enabling it to land and take off almost anywhere, sliding over ice or bumping on a dirt runway.

We've flown 300km over the jagged mountains of the Antarctic Peninsula (one of the fastest-warming places on Earth), which snakes up towards South America, and then out over the collapsing Wilkins Ice Shelf. Here, ice chunks bigger than city blocks lie entombed in a frozen sea.

There are no landing strips, just endless white. And the unexpected buzzer makes things a lot worse. Why did my Thomson Reuters video colleague Stuart McDill, sitting in the co-pilot's seat, and I accept this assignment? I look over at Canadian pilot Steve King, luckily he's not freaking out and looks reassuringly concentrated despite the alarm. A second or two after the buzzer sounds, the skis touch down gently on flat, slushy snow and ice and the plane slithers to a halt. He turns off the twin propellers, clicks a few buttons on the dashboard above his head and climbs out of the plane in his red jumpsuit and blinks into the hazy sunshine.

"Oh, didn't I tell you beforehand about the buzzer?" he chirpily asks as we passengers clamber down the ladder onto the bright ice. I'm still shaky when he tells us the alarm just before landing is a badge of honour for polar pilots – a warning that the plane has slowed to stall speed of about 105km per hour – meaning it is going to fall out of the sky. That sounds a daft goal for a pilot, but there is a safety logic. The sluggish speed meant the plane would travel a minimum distance before coming to a halt, reducing risks that it might fall into a hidden crevasse, the main worry for landing on unknown ice and snow. So, for King, the buzzer going off just as we landed was a stroke of aviation magic.

As the propellors spin to a stop, the silence takes over on a near-windless day by the Southern Ocean. No one has ever been to this spot before – and no one will ever come back. No one knew at the time in January 2009, but this part of the ice shelf was weeks away from breaking up for good in April into massive icebergs, probably after being in place for thousands of years. Now the region is open water in the brief Antarctic summers.

The scientists have come here because ice shelves, vast tongues of ice floating on the sea around much of Antarctica, hold huge clues to predicting sea level rise that threatens places from New York to Buenos Aires and coastlines from Pacific islands to Bangladesh. When we visit, the Wilkins Ice Shelf covers an area about the size of Jamaica.

McDill and I visited the British Antarctic Survey's Rothera Base for two weeks, a rare insight not only into the science but also the human side of life underpinning research in Antarctica. That landing was a fairly routine day for staff in a place where scientists expose themselves to risks – ranging from engine failures to crevasses – to help understand the world's biggest deep freeze.

Overall, Antarctica is the size of the US and Mexico combined and contains enough ice to raise world sea levels by about 58 metres if it ever all melted.[1] Scientists say there are more signs that Antarctica is thawing at the edges and, most alarmingly, that an 'irreversible' melt might already be under way.

Antarctica is divided into two main parts, East and West, each with its own risks. East Antarctica, the giant taking up most of the continent, has the coldest and most stable ice sheet, locking up more than 90 per cent of the ice. West Antarctica, which includes the Antarctic Peninsula and the Wilkins Ice Shelf, is the most vulnerable to a thaw.

A decade ago, the Wilkins was the focus of research but during the 2020s, scientists have shifted attention south, to the even more remote region of the Thwaites glacier. It's sometimes called the 'doomsday glacier', the widest glacier on earth at 120km across where it meets the ocean. The worry is that a collapse of the Thwaites could be the first domino to fall on the West Antarctic Ice Sheet, unleashing a flow of ice pent up inland that could lead to sea level rise of 3.3 metres[2] over centuries, or millennia. And the risks are growing – a bit like on our plane, the scientific alarm buzzer is ever shriller.

As previously noted, the UN's expert group, the IPCC, said in August 2021[3] that global sea levels are likely to rise by up to about a metre this century if man-made greenhouse gas emissions keep rising. The report includes dire warnings about the thaw under way in Greenland and Antarctica. Even if global warming is limited to the goals of Paris Agreement, it says "there is limited evidence that the Greenland and West Antarctic Ice Sheets will be lost almost completely and irreversibly over multiple millennia". Irreversible is, of course, planet changing.

On the worst trajectory of increasing greenhouse gas emissions, sea levels are likely to rise by between about two and seven metres by 2300, it said. That is above the upper bound of 5.4 metres considered in a 2019 report, based on a slightly different scenario. Radical action now to limit greenhouse gas emissions and promote a more sustainable economy could curb the rise in sea levels, in the best case, to less than a metre by 2300, the 2021 report says. But scientists are unable to predict with any degree of precision what will happen to ice sheets in a warming world. All the IPCC scenarios for rising oceans have huge ranges, especially beyond the next few decades.[4] Much of the increased risk seen by the IPCC this century is driven by Antarctica – every extra centimetre of meltwater makes floods more frequent in low-lying regions where tens of millions of people live. At the other end of the planet, Greenland is also pouring water into the ocean, threatening coasts and cities.

"Not keeping to the Paris climate agreement really commits cities to the sea," climate scientist Anders Levermann tells me. "And in 2300 sea level rise is not done. It's going to continue." He says that the world is still far from understanding the risks of what is happening in Antarctica. There may still be 'unknown unknowns' – unexpected processes scientists have yet to imagine – about ice sheets.

Under the 2015 Paris Agreement, governments promised to limit the rise in average global temperatures to "well below" 2°C above pre-industrial times while pursuing efforts for a 1.5°C maximum, but even such rises in temperatures could be the tripwire for a meltdown of ice sheets.

Sea levels are condemned to rise in the very long term even if governments meet the Paris targets, partly because of instabilities in Greenland and Antarctica, the IPCC said in 2021.[5] "Over the next 2,000 years, global mean sea level will rise by about 2 to 3 metres if warming is limited to 1.5°C," the IPCC said, and by 2–6 metres with 2°C of warming.[6] Bear in mind that temperatures are already up about 1.2°C.

David Vaughan, Director of Science at the British Antarctic Survey who led our flight to the Wilkins, says research is likely to stay in West Antarctica, especially to understand warmer ocean waters that are gnawing away at the ice from below. "The West Antarctic Ice Sheet – Thwaites, Pine island and a few other small glaciers – is where we should be working. They are the biggest uncertainty about sea level rise. I can't see that focus changing in the next few years," he told me in 2021.

He laughs when we reminisce over that day on the Wilkins – especially the risky timing of landing on a place that turned out to be about to disappear off the map.

After getting out of the Twin Otter plane on the Wilkins, Vaughan beams and starts jumping up and down on the slushy snow and ice, dancing for joy like a kid. "This is amazing!" he exults, a switch from the eminent professor we've got to know back at the scientific base. My colleague McDill jokes that it may be unwise to jump up and down on a collapsing ice shelf. But Vaughan, who has spent most of the flight studying the shattered ice out of the window, knows we're no more than ants on a skyscraper.

On the flight from the British Antarctic Survey's Rothera Base we've even passed over the eponymous Vaughan Inlet on the Antarctic Peninsula – the continent is so little known that places are sometimes named after people who are working here today. The Vaughan Inlet, honouring David's work on the break-up of Antarctica's ice, is near a place called Shiver Point.

From the sky, the Wilkins looks like a giant has angrily emptied out a jigsaw puzzle, his brain numbed by the jumble of white and bluish pieces, to create a senseless clutter. Or it's like someone has dropped a massive wedding cake and shattered icing is everywhere. There's no way of easily gauging the scale, there are no buildings or trees to give a sense of size in Antarctica, in fact there are only a few colours: black rocks, white ice and blue sky and water. Many of the chunks of ice we see are bigger than the Empire State building in New York or the Burj Khalifa in Dubai, lying on their sides. The only sense of perspective when we fly low has come from the small black dots on the sea ice – up closer we see they are seals in the sunshine on a windless day, craning their necks.

Vaughan explains to us that glaciers are giant rivers of ice, built up from snowfall. In permanently frozen regions like most of Antarctica, fresh snow weighs down on older layers below, crushing snowflakes and squeezing out air to form solid ice that becomes a glacier. Gravity draws the glacier downhill towards the sea – often meandering like a slow-motion river or icy serpent. When the glacier reaches the sea, it can break up or, in some cases, keep flowing outwards to form a tongue of ice floating on the water – an ice shelf like the Wilkins that can be hundreds of metres thick. At the outer edge of ice shelves, chunks eventually snap off into the sea as icebergs. It's all a natural process, but global warming is accelerating the slide. The Wilkins Ice Shelf juts about 20 metres above the sea with the rest under water – like an iceberg with only the tip showing.

The increasing worry among Antarctic scientists is that ice shelves act as natural brakes that hold back large amounts of ice inland, slowing the flow of the ice sheets towards the ocean. The Wilkins holds back ice equivalent to about 1cm of sea level rise if it all melted, tiny compared to the Thwaites.

Glaciologists have come up with an array of imaginative analogies to explain what's going on with ice shelves. In the worst case, some glaciologists fear Antarctica's ice, which is 4,776 metres deep at its thickest point,[7] is like a gigantic wine bottle lying on its side, with the ice shelves acting as the cork. When the cork fits tight, everything's fine. Remove the cork, however, and almost immediately the whole bottle starts to gurgle out.

Others compare Antarctica to a Gothic cathedral – a building whose weight squeezes the walls outwards and needs extra support from the sides, or flying buttresses, to prevent collapse. Richard Alley, a professor of geosciences at Pennsylvania State University, told a US Congressional committee in 2019:[8] "Some early Gothic cathedrals suffered from the 'spreading-pile' problem, in which the sides tended to bulge out while the roof sagged down, with potentially unpleasant consequences. The beautiful solution was the flying buttress, which transfers some of the spreading tendency to the strong earth beyond the cathedral. Ice sheets also have 'flying buttresses', called ice shelves."

Less poetically, he and others also compare the Antarctic ice to a lump of pancake batter – when it's cold, it's a stable lump. Heat it in a frying pan and it will quickly splay out.

Among other explanations, the Earth Institute at Columbia University says that land-bound glaciers behind ice shelves are like people jostling for space. Glaciers "are constantly pushing seaward. But because many shelves are largely confined within expansive bays and gulfs, they are compressed from the sides and slow the glaciers' march – somewhat like a person in a narrow hallway bracing their arms against the walls to slow someone trying to push past them".[9]

Ice shelves themselves don't add to sea level rise when they break up – they are already part of the ocean floating on the water. But it is their effect in releasing the pent-up ice behind them that can make a massive difference. Since Antarctica's ice is so vast, even small changes around the edges need to be monitored because they can affect coasts around the world. Relatively speaking, the ice here dwarfs mountain glaciers from the Andes to the Himalayas. In addition, about 75 per cent of Antarctica's coast has ice shelves[10] extending offshore – the largest, the Ronne-Filchner, covers an area a bit smaller than Spain. Other large shelves are Ross and McMurdo. So that's why we flew to the Wilkins for a first-hand view.

In the early 1990s, the Wilkins Ice Shelf covered about 17,400 sq km, according to the US National Snow and Ice Data Center.[11] At the time of our visit, it's shrunk to 10,300 sq km, after repeated collapses in a warming world. Large parts of eight other smaller shelves around the northern part of the Antarctic Peninsula have already broken up.

By the time we are there on the Wilkins, the British Antarctic Survey reckons that a total of 25,000 sq km of ice shelf has been lost from around the Antarctic Peninsula since the 1950s. That lost area is bigger than Israel or roughly the same size as the US state of Vermont.

It's January 2009 and just past midsummer in the southern hemisphere – the spot we've just landed on is a 40km long 'ice bridge' of the Wilkins Ice Shelf connecting mainland Antarctica to Charcot Island offshore to the west. Break-ups of the shelf mean this ice bridge is about 500 metres wide in 2009 at the narrowest, down from tens of kilometres in the early 1990s. In satellite pictures at the time, the Wilkins ice bridge looked like a jagged sliver from Charcot Island pinning the bulk of the ice shelf against the coast of Antarctica. A year earlier, in 2008, the British Antarctic Survey flew down here, without landing, and took dramatic pictures of the thinning ice bridge that were broadcast worldwide. It concluded that the Wilkins was "hanging by a thread".[12]

My colleague McDill sets up his tripod and camera on the ice and Vaughan steps up. "We've come to the Wilkins Ice Shelf to see its final death throes," Vaughan says. "It really could go at any minute." After the 2008 British Antarctic Survey flight, "miraculously we've come back a summer later and it's still here. If it was hanging by a thread last year, it's hanging by a filament this year."

He tells everyone that we shouldn't linger – and correctly predicts that the shelf is likely to break up in coming weeks or months. As part of the trip, Vaughan and colleagues slot together what becomes a 4-metre pole topped by a GPS satellite transmitter to be stuck into the ice for a Dutch-led experiment to detect movements in the shelf. It will give hints to movements that could herald a collapse, caused by winds or ocean currents. Having assembled the pole, Vaughan pulls out a tiny strip of paper that will connect the battery and activate the GPS. He does so and the GPS pole starts bleeping.

Good, it seems to be working. But after about a minute, the bleeps abruptly stop and there's no other sign of life from the pole. This starts a debate among the scientists: yes, it's probably working perfectly and was programmed to bleep only for a few seconds to show it was okay. But what if the battery has slipped out of place, or the GPS is broken and this part of the trip is in vain? No one knows – there are no written instructions, and no mobile phone, satellite phone or radio connection to check with the Dutch scientists back at the Rothera Base 300km away. The scientists do the usual things when something like a TV remote control doesn't work back home – shake it, take the battery out and put it in again, knock it a couple of times. It doesn't make a difference. The beeps don't resume. Vaughan rightly concludes that it's working – endless bleeping would be a pointless waste of energy with only the odd passing Antarctic skua – an often-aggressive, seagull-sized brown bird – likely ever to hear it.

While on the ice, King, the pilot, advises passengers to stay near the plane – walking anywhere is a risk because there might be treacherous fissures.

McDill does a memorable piece to camera about landing in a spot never visited before and we all take photos to capture the moment. Afterwards, however, they look unremarkable – the background is just a flat, continuous white of snow beneath a blue sky mottled with clouds, all as if taken on an empty plain – no ice cliffs, no sea, no drama.

King shepherds everyone back aboard, again warning that you never know when this place might shatter. We take off and he flies nerve-janglingly low, skimming over the frozen sea to give McDill a view for taking video – sometimes even below the flat top of the ice shelf, which is only about 20 metres above the ocean. And a bit later, flying along a several kilometre-long crack between two massive chunks of the ice shelf, the wings seem to be almost touching the sides.

McDill films from the co-pilot's seat beside King – I'm impressed his nerves are steady enough. Viewing it later, scientists say it's like the Star Wars movie *A New Hope* when Luke Skywalker skims low above the Death Star.

"Just another day cheatin' death," King remarks wryly at one point.

The Twin Otter, a Canadian-built de Havilland plane, is a workhorse in Antarctica. With a wingspan of about 20 metres and an ability to accelerate and take off within about 400 metres, every kilo counts. No one takes flying lightly – danger is everywhere in Antarctica.

Among other ice shelves that have broken up on the Antarctic Peninsula are the George VI – named after the late British King and father of Queen Elizabeth – or the Prince Gustav, named after the man who became king of Sweden in 1907. Antarctica's names, a snapshot of history in the early twentieth century, underscore the isolation of the continent, and the difficulty of knowing how it might melt.

The continent was sighted[13] in 1820, by Russian, British and American expeditions, meaning many names date from around then, with a mixture of royalty and explorers and competing claims – Argentina, Britain and Chile all have overlapping stakes to territory on the Antarctic Peninsula, 'frozen' under the 1959 Antarctic Treaty which says the continent belongs to no one. An era of exploration followed the early whaling pioneers. Norwegian Roald Amundsen was first to reach the South Pole in December 1911, a month before Briton Robert Falcon Scott and four companions. Scott and his men, trapped for days in a tent in a blizzard and suffering from hunger, exhaustion and frostbite, died on the return journey.

The Wilkins itself is named after Australian George Hubert Wilkins, an early Antarctic aviator and explorer of both poles. A year after he died, aged 70 in 1958, the US Navy scattered his ashes at the North Pole.

The problem for understanding the risks is that Antarctica's thaw is seen by so few. It's hard to grasp one of the most alarming effects of man-made global warming caused by rising greenhouse gas emissions from cars, factories and power plants when it is thousands of kilometres away. Similarly, the historic nature and the sense of such giant markers in the landscape being immutable makes the concept of change hard to fathom.

Ice shelves have indeed been around a long time. A study of ocean sediments beneath one of the collapsed Larsen ice shelves on the Antarctic Peninsula, for instance, found no recent traces in the seabed of algae – tiny marine plants – that require sunlight to grow. That indicated that the ice had been in place, blocking sunlight, for at least 10,000 years. Those Larsen shelves are named after a Norwegian whaler who sailed perilously far south in the late nineteenth century.

But the loss of ice shelves is already changing the maps of Antarctica. After taking off from the Wilkins, we fly with the Twin Otter for three more stops to allow British scientist Alison Cook to update maps made in 1975 by the US Geological Survey, before the satellite and GPS age allowed far greater precision. She's checking and updating GPS coordinates – and also doing her bit to reduce a gender imbalance in Antarctic science where work on the continent has been mostly done by men.

On one flight, we fly towards a 'nunatak' – a Greenlandic word for a lump of rock sticking out of the ice – in the Wilkins Ice Shelf known as Merger where Cook can take GPS measurements. For this flight, I get to sit in the co-pilot's seat and even hold the controls for a few minutes while King does paperwork, all the while keeping a wary eye on me. Within minutes another buzzer sounds. I jump.

"The plane reckons we're too close to that cliff over there," King says nonchalantly, pointing towards a steep mountainside looming alongside. He flips a switch to turn off the bleeper. King lands on yet another trackless spot of ice, after circling to check for crevasses.

Cook has details of how a US Geological Survey team got to this nunatak with dogsleds over the Wilkins in 1975. Dogs have been banned from Antarctica since 1994 as part of a treaty to protect the environment. Mapping has been plagued by gaps. "Strange as it may seem, the far side of the Moon is better mapped than parts of our own planet. Position errors of more than 100km were found in 1975 on the most up-to-date maps published of one part of the Antarctic," a US Geological Survey report found.[14]

It's hard to imagine the effort those US scientists went to, compared to just zipping in and landing with a plane as we did. Cook has a copy of the scientists' notes from 1975 which warn of "heavy crevassing, difficult crossing from Alexandra Island", with details about where the scientists had to camp for the night. She says we're lucky to be able to stay just a couple of hours and fly back to a warm base.

The US scientists who were last here left a cairn about 3ft high to mark a metal spike in the ground, the highest point in the region. Cook and we others dismantle the cairn – it's odd to think that human hands once touched each of these stones – so that Cook can take a new more precise reading from the spot with a GPS mapping device. Weirdly, underneath the cairn the scientists have left a can of butter, leaking a whitish goo through the rusty sides. We've no idea why – perhaps a gift if a future expedition ran out of food? On the highest point of the hill, the cairn is visible from far across the ice shelf. If I was starving, I'd probably try decades-old butter. But not today – it's sunny, the temperature is above freezing and we've plenty of food for picnics. We leave the tin in place.

In the weeks after our trip to the Wilkins, scientists keep monitoring the website of the European Space Agency, which has a satellite taking pictures of Antarctica. I also check the map every day – sometimes clouds mean there's no clear view of the Wilkins, other days show no change, some days are missing. But on 4 April, a Saturday, the grainy photo shows the ice bridge shot through by cracks. I call Vaughan late in the evening, who confirms the ice bridge had just shattered.

It has changed the map of the continent. "Charcot island is a true island for the first time in recorded history," Vaughan says. What was once an ice bridge connecting the island to the mainland of Antarctica is now a strait with open water.

Two days later, Hillary Clinton, then US Secretary of State, made a plea for action on climate change in a speech in Washington. "With the collapse of an ice bridge that holds in place the Wilkins Ice Shelf, we are reminded that global warming has already had enormous effects on our planet, and we have no time to lose in tackling this crisis," she tells an international meeting about the Antarctic and the Arctic.[15]

Since then, more of the Wilkins ice has washed out to sea.

In early April 2009, the final ice-bridge collapse corresponded to an area loss of 330 sq km, according to a 2017 study by scientists based in Germany.[16] Ever more cracks have appeared and other icebergs have broken off since then. Most stories about climate change describe slow-moving trends, and too often these pass under editors' radar. But the Wilkins collapse was breaking 'news' of a type likely to become ever more frequent this century.

Now, figuring out exactly what is happening at Thwaites is the main puzzle. Ice from this vast glacier splays out into the Amundsen

Bay, unchecked by mountains, nunataks or other obstacles that can act as natural brakes.

In 2014, glaciologist Eric Rignot, of the University of California, Irvine, and NASA's Jet Propulsion Laboratory in Pasadena, California, and colleagues were among the first to suggest that an irreversible melt[17] had begun in the region. Their conclusion was partly based on satellite radar observations of Thwaites and other glaciers in the region, including Pine Island, Haynes, Pope, Smith and Kohler.

Rignot gave a stark assessment of the situation in a 2019 statement to back a lawsuit by a group of young Americans seeking far tougher climate action by the US government. He declared that man-made greenhouse gases have created "a ticking time bomb for the planet's ice sheets. Some of our ice sheets are already in an unstoppable melt and disintegration. ... Between the irreversible melting of portions of Greenland's and Antarctica's ice sheets, humanity has already committed itself to a 3–6-metre rise in sea level."[18]

Bigger than Florida and almost the size of Britain, Thwaites gets its apocalyptic reputation as a doomsday glacier because it is melting ever faster as ocean waters gnaw away at its underbelly. According to the International Thwaites Glacier Collaboration (ITGC), "Over the past 30 years, the overall rate of ice loss from Thwaites and its neighbouring glaciers has increased more than five-fold. Already, ice draining from Thwaites into the Amundsen Sea accounts for about four percent of global sea-level rise."[19]

The ITGC, launched in 2018, is a $50 million project run by the US and Britain to deploy scientific instruments on the ice. It originally got its name because of a scientific disagreement about how to spell on different sides of the Atlantic. The US favoured the 'International Thwaites Glacier Program', while Britain favoured 'The International Thwaites Glacier Programme'. (In British English, 'program' is often limited to computing.) Vaughan recalled that he had jokingly suggested a tie-breaking compromise of the 'International Thwaites Glacier Program(me)'. But that looked odd to everyone, so they struck on 'Collaboration'.

The Thwaites is remote, even by Antarctic standards – it's about 1,600km to Rothera, and 2,000km to the US McMurdo base in the other direction around the coast. "Antarctica is already a difficult place to get to ... It's going to take both nations' capabilities in Antarctica in order to set all of these studies out on the ice," says Ted Scambos, a senior research scientist at the US National Snow and Ice Data Center in a video about the ITGC.[20]

The Thwaites alone could raise sea levels by 65cm if it all collapsed, unlocking a wider flow into the ocean of the West Antarctic Ice Sheet.[21] That vulnerability is largely because much of the Thwaites rests on bedrock far below sea level and warmer water currents are seeping ever deeper under the ice shelf, floating on the ocean. Much of West Antarctica would be a deep open ocean if the ice vanished. Maps of Antarctica imagining no ice show a patchwork of islands in this region. And the melt of the Thwaites makes the end of the glacier ever more vulnerable to cracking. In turn, a retreat of the end of the glacier may expose ever taller ice cliffs that risk collapsing under their own weight, accelerating losses.

Thwaites glacier could "potentially disintegrate this century", destabilising the entire West Antarctic Ice shelf, the IPCC said in 2021. Thwaites could be "the ideal site for monitoring early warning signals of accelerated sea level rise from Antarctica. Such signals could possibly be observed within the next few decades," it said.

One of the fears in the scientific community is that rising air temperatures could lead to pools of meltwater on the surface of ice shelves around Antarctica, most of which are now frozen even in summertime. A 2020 study estimated that between 50 per cent and 70 per cent of ice shelves buttressing glaciers inland might be vulnerable to the pools of meltwater that could then flow down into cracks tens or hundreds of metres deep, refreeze and then shatter the ice – a process known as hydrofracturing.[22]

"The ice shelves – that's the weak spot, where the atmosphere, the ice and ocean interact," the study's co-author Jonathan Kingslake, a glaciologist at Columbia University's Lamont-Doherty Earth

Observatory, said in a statement.[23] "If they fill up with meltwater, things can happen very quickly after that, and there could be major consequences for sea levels."

Worries about such 'hydrofracturing' breaking up ice shelves have added to fears for sea level rise from Antarctica. Robert DeConto, of the University of Massachusetts Amherst, and David Pollard of Pennsylvania State University say hydrofracturing could expose ice cliffs perhaps 100 metres tall that would start to collapse under their own weight, releasing vast chunks of ice splashing into the ocean and speeding the flow of glaciers towards the sea.

Their 2016 findings set off alarm bells among scientists by saying that Antarctica could contribute more than a metre to sea level rise by 2100, and perhaps more than 15 metres by 2500 if greenhouse gases continue to rise and the ice breaks up.

That was far more than expected and many scientists were sceptical about widespread "marine ice cliff instability" in Antarctica. The August 2021 IPCC report marks acceptance of the theory as plausible, but a remote possibility. As earlier noted, an IPCC graph of projected sea level rise until 2300 has a note saying that a rise of more than 15 metres "cannot be ruled out".

Robert Kopp, a professor in the Department of Earth and Planetary Sciences at Rutgers University-New Brunswick and an IPCC author, stressed there are still huge unknowns.

"We have low confidence in all sea level projections on these timescales," he told me of the IPCC findings.

Still, Vaughan reveals that there are already signs that cliffs on the front of Thwaites glacier are getting so tall that they are splintering off under their own weight. On parts of the Thwaites, "Icebergs are forming because the ice cliff is so high that it cannot support its own weight. It sort of flakes off ... It's like a wooden pit prop in a mine, and you squeeze it until it buckles."

And there are other quirks along the way. The ice sheets on Antarctica and Greenland are so huge that they exert a gravitational pull on the sea nearby, pulling it up. If the ice melts, the tug of gravity will lessen and the sea around Antarctica and Greenland will fall, while raising the ocean water on the far side of the world. A melt of Greenland, for instance, will have the biggest impact on raising sea levels in the South Pacific.

The Thwaites Collaboration has launched a number of projects to figure out what is happening, and these are often given names based on memorable but awkward acronyms, such as PROPHET, DOMINOS, GHOST, THOR, MELT and TARSAN. It's a bit of a stretch to tease the letters of PROPHET from the formal title, but they are clearer with upper case clues: 'PROcesses, drivers, Prediction: modeling the History and Evolution of Thwaites'.[24] TARSAN is similarly hard and they probably struggled in vain to find a 'Z' instead of an 'S': 'Thwaites-Amundsen Regional Survey And Network integrating atmosphere-ice-ocean processes'.[25]

Inland on the ice, the MELT team drilled a 600-metre hole about 35cm across through which they lowered a robot submarine to test the waters below near the grounding line, the place where the underside of the glacier rests on bedrock.[26] Just to bore the hole they had to use tonnes of fuel to melt water that was placed in a large rubber container, a bit like a small swimming pool. Scientists also had to be careful to avoid slipping into the vertical hole, and to ensure it did not simply re-freeze before they could lower their Icefin robot submarine down the hole.

Icefin, equipped with sensors and four cameras, found the water beneath the glacier was about 2°C above freezing, the first observation of temperatures at the critical grounding line point where the bottom of the ice touches the seabed beneath the Thwaites. Coincidentally, it also spotted small fish and an anemone.

Scientists say there's a worrying hint of an ancient precedent for a collapse of West Antarctica – in the Eemian Period, about 130,000 to 115,000 years ago when natural shifts in the Earth's orbit made it about as warm as now, sea levels were between 6 and 9 metres higher than they are currently. Scientists don't think that all that ancient melt can have come only from Greenland, which locks up ice equivalent to only 7 metres of sea level, so they think that West Antarctica was at least partly the culprit.

By contrast with West Antarctica, East Antarctica is far colder and rests mainly on land that is not as vulnerable to less chilly seas. The South Pole is here on an icy plateau.

Among other Thwaites projects, in THOR – THwaites Offshore Research[27] – scientists based on a ship have been drilling into the seabed off the end of the glacier. By examining layers of silty sediments built up over centuries they can figure out how the glacier has grown or contracted through time. Big stones in the seabed were likely dropped into place from passing icebergs in which they were trapped, which also means the region was not blanketed by an ice sheet at that time.

Scientists have to be inventive in examining the gooey slime they extract, according to geologist Linda Welzenbach, who wrote a blog aboard the *Nathaniel B. Palmer* research ship as part of THOR. She reflected, "The sediment cores we are collecting consist of soft mud and often hold large rocks and pebbles, so handling them involves tools you wouldn't expect – dry wall spatulas, kitchen sponges, water buckets, kitchen spoons and metal – tools that aren't too different from what you find in a dentist's office."[28]

The Thwaites, unstable and with ice pinned to the seabed, is almost as scary as anything dreamt up in Hollywood. After all, the 2004 disaster movie *The Day After Tomorrow* depicts an icy horror story in which a yawning crack opens through a US research camp on the Larsen B ice shelf on the other side of the Antarctic Peninsula from the Wilkins. The movie then goes on to a head-spinning mix of cataclysmic storms, rising seas and plunging temperatures.

Back in the real world, Antarctica attracts some of the world's top scientists, often trying to figure out sea level rise, as well as studying topics such as geology, penguins, whales and krill, permafrost or how the underside of icebergs scour the seabed.

The science teams are backed by workers including plumbers, carpenters, electricians, cooks and medical staff at bases around the continent. More than 100 people live in Rothera during the summer months, with about twenty-two 'winterers' staying on year-round to keep the base ticking over and the pipes from bursting. Some have the longest contracts in Antarctica and have spent up to two and a half years away from home and have developed extraordinary skills in the dark months. On arrival at Rothera, one of the first bits of advice, before the safety briefing, was: "don't play pool with a winterer". They're simply too good.

When we visited, meteorologist Rob Webster had learnt to play the drums while on a two-and-a-half-year contract as part of a Rothera group appropriately dubbed 'Nunatak'.[29] You may think what's so special about learning to play an instrument? But his first gig drew a global audience of hundreds of millions of people – Nunatak was Antarctica's contribution to a 'Live Earth' concert beamed around the world from every continent, organised by former US Vice President Al Gore in 2007. When we meet, we joke that he was probably more successful on his debut than any artist in history. Webster's departure coincided with our flight back to Chile. He decided not to come to a restaurant with us, preferring to sit outside in a square in Punta Arenas to watch the crowds on a warm evening and eat mangoes – experiences unavailable in Antarctica.

Before coming to Rothera, everyone has had to learn emergency first aid, including some basic surgery techniques, going as far as draining a punctured lung by stabbing a hole between the ribs. Everyone has practised putting up tents in the freezing cold and learnt how to send an SOS by radio in case of a plane crash.

Antarctica can be beguilingly calm – when we visited Rothera in the Antarctic summer temperatures were warmer than the winter in Cambridge, the headquarters of the British Antarctic Survey (BAS). Sometimes the temperatures were about 7°C.

Athena Dinar, the communications manager for BAS, who accompanied McDill and me in Rothera, with a tireless good humour for daft questions, took us on our first day for a walk along the coast by Rothera, out along the rocks which were playing host to a few seals. Away from the base, we stopped and she told us to stay still.

Here, by the shore, she tells us to listen to the very sound of global warming – the fizz as air trapped in icebergs floating in the sea bubbles out into the open air from the water. It's an entrancing hissing, popping sound that you have to strain to hear. And this faint icy symphony is a memorial to the ice: the air released in the bubbles is often hundreds of years old, from a time when carbon dioxide concentrations were much lower than the levels now.

Towards the end of our visit, the base throws a party – including an ancient block of ice for gin and tonic recovered from the bay that may be more than 1,000 years old. Vaughan tells us he knows it's old because it is bubble-free and transparent – almost all the air has been squeezed out by the pressure of the ice above when this chunk was part of a glacier.

At Rothera, some of the solutions to making life comfortable are tried and tested, rather than high tech. Visitors have to learn how to put up a tent and operate a primus stove. McDill and I had a dinner of sardines from a tin marked "best before 2004" – five years before our visit. British scientists have tended to use lower tech wooden boxes and other materials than American expeditions. Wood doesn't rot in Antarctica because of the cold, and in the minds of British explorers, is more durable than plastic.

In a tent in Antarctica, you can have a dozen layers separating you from the chill, not counting any clothes you might have on. At the bottom, the tent has a liner that goes on the snow. Above that you lay out a plastic groundsheet, then a wooden board. After that comes a foam mat, an inflatable mattress, a sheepskin rug, a green waterproof sheet, a fleece blanket, a sleeping bag and a cotton sheet to line the sleeping bag.

♦♦♦

From the Wilkins to the Thwaites, evidence for a drastic change in Antarctica is piling up.

Until several years into this century, many scientists reckoned that Antarctica could stay frozen, even with strong levels of man-made global warming. The idea was that it is a giant deep freeze, simply too cold to melt no matter how severe global warming becomes. The highest temperature ever recorded at the South Pole, one of the coldest parts of the continent, is -12.3°C in 2011. The coldest temperature recorded by a weather station was -89°C at Russia's Vostok Station in Antarctica in July 1983. In the north, the Argentine research base Esperanza on the Antarctic Peninsula recorded a temperature of 18.3°C in February 2020, the continent's highest on record.[30]

Earlier this century, the consensus by the IPCC's panel of scientists was that Antarctica could reduce sea level rise this century, in apparent defiance of global warming. That idea was founded on the concept that warming will mean more water evaporating from the seas into the air – and when these more humid winds blow over the frozen continent some of the moisture falls as snow, making the ice sheet bigger. As recently as 2007, an assessment by the IPCC noted that computer projections this century "indicate that the Antarctic Ice Sheet will receive increased snowfall without experiencing substantial surface melting, thus gaining mass and contributing negatively to sea level".[31]

That's all since gone into reverse.

Antarctica is no longer seen as a potential benign extra store of ice to offset sea level rise even though scientists say billions of tonnes of more snow are, in fact, falling every year. The continent is simply thawing faster than the extra snow accumulates.

There are still many doubters about Antarctic science in the world, especially US Republicans, even though a lot of the painstaking evidence of shrinking ice and rising seas has been compiled by American scientists. Former US President Donald Trump caused a mixture of bewilderment and outrage in 2018 by saying in a television interview, "The ice caps were going to melt. They were going to be gone by now, but now they're setting records, okay? They're at a record level."[32]

It's unclear what he meant by "ice caps" – in scientific terms, ice caps are areas of ice on land covering less than 50,000 sq km,[33] anything bigger is an ice sheet. And ice is shrinking pretty much everywhere. Even sea ice floating around Antarctica, which had seemed to expand in defiance of the warming trend, set a record low in satellite records in 2017, according to the US National Snow and Ice Data Center.[34]

And there are still massive gaps in understanding Antarctica.

That's partly because there are so few observations – there are written records of Greenland, for instance, dating back to Viking times, centuries before the first confirmed sightings of Antarctica. And satellite observations were only possible starting in the late 1970s.

One of the mysteries of our visit to the Wilkins was about the GPS monitor Vaughan installed on the ice. The Dutch scientists found that, after the ice bridge shattered, the GPS kept transmitting, apparently still intact on a vast iceberg formed in the break-up of the ice bridge. Its signals, recorded by satellites, trace a series of pirouettes along the coast of Antarctica before it either fell in the sea or ran out of power about a year later.

And, like the Wilkins, it's now gone from the face of the Earth.

2

IN FIJI: ONE MAN, THREE HOMES

Vunidogoloa: the first village in the world to begin relocating to higher ground due to sea level rise.

Britain's Prince Harry, on a visit to Fiji in 2018

Sea level rise is already forcing some remote communities to move inland, such as in Fiji in the Pacific Ocean where the government relocated 150 people in the village of Vunidogoloa to higher ground in 2014 from a flood-prone coast. It was a wrenching shift for families who had to abandon graves of their ancestors, but it is generally hailed as a success that could guide displacement elsewhere.

"THIS IS MY HOUSE," THE chief of a Fijian village says, pointing to an old lump of concrete and a few wooden posts jutting out of the brownish sand of a beach shaded by coconut palms. This is all that remains of the boyhood home of Simione Botu, head of Vunidogoloa village in eastern Fiji. He has already moved inland – not once, but twice – to escape worsening floods along the coast. "Our heart is here. Our forefathers were here," he tells me, standing on the narrow beach by the rotted foundations of his first home, washed away decades ago by the Pacific Ocean.

During a 2018 visit to Fiji, in the South Pacific east of Australia, Britain's Prince Harry mentioned Vunidogoloa village as the first to move inland anywhere in the world because of sea level rise. It is a dismal title that could be competed over by several other communities around the globe.

Botu himself is one of a vanishingly small group of people anywhere who have been forced to move from two homes because of coastal flooding. His trail of displacement is a tale of stubborn resilience, with a happier conclusion in his modern, third house up on the hillside, built by the villagers and the Fijian government.

Old Vunidogoloa village, where Botu grew up, lies on a coast exposed to storm surges caused by Pacific cyclones. It is at the inner end of Natewa Bay, which can act as a funnel for water blown from the ocean into the narrowing inlet. The village is also at the mouth of a meandering river that capriciously devours the shoreline, compounding the harm of sea level rise.

Dressed in a blue T-shirt and black shorts, Botu is certain when I ask him what happened to make his house vanish. "Climate change," he replies. "Sea level rise, floods, caused this. So, we have to shift to the new site."

"No matter what we did, the water came through the village," Botu said of repeated attempts to hold back the sea and the estuary snaking through the village, with makeshift walls. Sea water intrusion disrupted agriculture and the cultivation of coconut, breadfruit and banana trees.

After years of planning, the entire village of 150 people moved to New Vunidogoloa, about 1.5km inland on a hillside owned by the villagers. The new village comprises thirty-three new one-storey wooden homes, painted light blue. The new village is widely hailed as a successful project since it opened in 2014, despite grumblings on both sides. The villagers accuse the government of failing to keep promises, while politicians who championed the project complain about ingratitude.

Fiji's government reported in 2014 that it had spent almost a million Fijian dollars (about US $500,000) to relocate Vunidogoloa, including building homes, fish ponds and a processor for helping to produce coconut oil. And Fiji plans to move more than forty communities away from the coast, setting an example for vulnerable nations in managing rising sea levels in the twenty-first century.[1]

Fiji is often a leader in climate policies – it was the first nation anywhere to formally ratify the Paris Agreement. And it is among those least responsible for climate change – Fiji emits a miniscule 0.006 per cent of global greenhouse gases.[2]

The move from Old Vunidogoloa, where only a few buildings are still standing which are slowly being overgrown by vegetation, has attracted praise as governments around the world worry about how to handle an accelerating rise in sea levels driven by melting ice from Greenland to Antarctica. However, when hearing about places like Vunidogoloa, a natural reaction is that: *Seas are rising around the world. What's so special about Fiji's coastline? Why isn't everyone moving inland everywhere if things are so bad?*

For climate sceptics, that question is often a sneaky foot in the door to raise doubts on how far sea level rise is a factor, or whether it's even happening at all. From West Africa to Alaska in the US, scientists say a nasty cocktail of factors is forcing relocations from the coast. In some cases, it's down to natural subsidence and other changes that have been happening for centuries but in others it's down to us – a blend of more powerful storms, shifting ocean currents and sea level rise linked to human greenhouse gas emissions.

In cases like Vunidogoloa, sea level rise is the proverbial last straw – or perhaps bale of straw – that breaks the camel's back and makes a vulnerable place uninhabitable. Sea level rise comes on top of flooding caused by more rain and more powerful storms that may also be stoked by greenhouse gases. And sea level rise will get steadily worse this century.

And it's not just a few remote villages at risk. Worldwide, approximately 680 million people, about 10 per cent of the global population, live in coastal areas less than 10 metres above sea level, according to the UN panel of climate scientists. The IPCC projects the number of people living in this low coastal zone could reach more than 1 billion by 2050.[3]

Fiji has been among the pioneers in figuring out what to do, starting right here on the beach in Vunidogoloa, on the island of Vanua Levu.

Botu laments the ruins of his first home as a few birds squawk from the dense vegetation behind the beach, apparently upset by our intrusion on what has become their territory since it was abandoned by the former villagers in 2014. The remains of Botu's house are mostly buried in the sand, covered with dead orange and brown leaves from trees nearby. The sand on this lower part of the small beach is dark, soaked by waves at high tide.

Botu is wistful for the past – this is where he played as a boy and went fishing in the 1960s, by the shore with warm soft sand – a place where he said he expected to live his whole life. He gestures out over the calm sea, an idyllic view across Natewa Bay with a few rain clouds above the green hills in the distance. Just along the beach he helps a woman who is deftly repairing a raft traditionally used for transport, made of long bamboo poles lashed together with string. Such rafts were sometimes used to get around the old village when it flooded, often with high tides at the time of a full moon.

After his childhood home became waterlogged, Botu built a new wooden house about 50 metres inland. He leads me through waist-high vegetation away from the beach, past an abandoned building with broken glass panes, to the one-storey wooden building. He says there are no nasty spiders or snakes here but I'm glad he's leading. This, his second home, is an attractive light yellow with a red door and a corrugated iron roof. The paint is cracking but at first glance the building doesn't look too rundown. Indoors, it's still used to store some building materials and wood. Outside, however, Botu bends down to show me the foundations, the wooden stilts on which the corners of the house rest, about 50cm above the ground. "This is the flood mark," he says, slapping the top of a wooden pile scarred by sea water. It came "right up to the top here. It nearly went to the floor." A seawater flood that soaks the floor can buckle the building and make it uninhabitable, he tells me. And the repeated floods meant the ground was unstable.

The village's problems were decades in the making. Born in 1957, Botu recalls how his parents' generation decided already in the 1950s to move from the shoreline – long before scientists detected man-made global warming or global sea level rise. Even then, the unpredictable river and the storms made it a risky site.

But there was widespread opposition.

"The first time they decided to move, old men from the coastal area came and told our forefathers, 'You should not move. We should do something to protect the village.' So, they made a sea wall of stone," Botu recounts. But "when the flood came, it went through" despite repeated attempts to reinforce the walls. "That's why they decided to move – people are afraid of that, sea level rise, floods, soil erosion, crops damaged." It was an agonising choice, partly because Old Vunidogoloa has a cemetery where the village families are buried. Villagers felt they had to weigh the interests of the living against respect for the dead.

Government officials came repeatedly to inspect the site and give advice ranging from how to build new sea walls to how to prevent flooding of latrines. The officials concluded that walls were costing the state too much and said, "look for a new site", according to Botu. Eventually, he said the villagers dropped their resistance when the officials bluntly asked, "What is more important to you: staying in your old site, or the life of the village?"

"That's why we decided to move, to save the life of the people," he said. In 2006, the villagers sought and went on to win aid from the Fijian government to relocate the village, a process that began in 2012.

Botu, who has four children and four grandchildren, now lives in one of these new cyclone-proofed homes – an airy single room with colourful decorations, and curtains to divide the space. There are problems, however. Indoor kitchens were not included, as expected by the villagers.

Outside his home, the village is lively as the early afternoon sun starts to ease down the sky. Four children laugh as they play a favourite game with old car tyres – one heaves a tyre to make it roll as far as possible while rivals try to hurl tyres from the side to knock it over. Under the shade of a mango tree, a group of women are sewing and cutting leaves to extract the woody stems to make brooms for sweeping the floors. Further up the hill, workers are tending a plantation of pineapples. Botu gets one and shares it with me – it is deliciously refreshing.

"People are happy living here, but it's not really what we are wanting," Botu says, adding the villagers still need support to fund extra improvements.

The villagers earn money from selling products including timber, sand, gravel, coconuts and grog – a drink made from the roots of the kava plant served in a coconut shell that gives a sedative, mildly euphoric feeling. The other food and fish they harvest is for local consumption.

Atop the hillside, the villagers are building a Methodist church. New Vunidogoloa is also called 'Kenani', the local version of 'Canaan'. In the Bible's Book of Exodus, Moses led the Israelites from slavery toward the "promised land" of Canaan, "flowing with milk and honey". Anchoring the move to the Bible has helped add a powerful motive to relocate. Botu reckons the church, a locally funded project separate from the relocation, will cost 250,000 Fijian dollars (US $125,000).

Among other advantages, the new village is by a gravel road where buses stop on the way to the towns of Labasa, the biggest on the island, and Savusavu. That means the locals have a far shorter distance to walk if they shop in town, and anyone who is ill can get to the doctor more easily. "From the old site it is very hard to go to town. There is no road – you had to walk from the old village to the main road, it took about 20–25 minutes to get to the bus," Botu remembers. Also, bringing food from the plantation down the hillside to the old village was a slog. "Now it's very easy for us. We just walk here to the main road and wait for the bus. Our plantation is just here," he says, pointing across the hillside.

Fiji's government, in a plan for fighting climate change sent to the United Nations in December 2020,[4] says that Vunidogoloa was the first of six 'climate vulnerable communities' partially or completely moved so far. New Vunidogoloa, with its solar panels for electricity, was a showcase at UN negotiations on climate change in Bonn,[5] when Fiji became the first small island developing state to lead the annual talks in 2017. Fiji's Prime Minister Frank Bainimarama has cast Vunidogoloa squarely as a victim of sea level rise – effectively pinning the blame on industrial greenhouse gas emissions from nations such as the US, the European Union, China, India and Russia.

"Today, we launch the first project in Fiji to save an entire village from the rise in sea levels caused by climate change," Bainimarama said in a ceremony in January 2014, opening the new village. "It is real. It is happening now."[6]

A few years later, Britain's Prince Harry made a far wider claim[7] during a visit to Fiji as part of a trip with his wife Meghan that also took in Australia, New Zealand and Tonga. "Just six years ago, Fiji's Vunidogoloa became the first village in the world to begin relocating to higher ground due to sea level rise ... This country is highly susceptible to the impacts of climate change. And it is having a profound effect on peoples' lives," he said. "We cannot ignore the reality of what is happening around us." Posters welcoming Harry and Meghan, with photos of both, were still up in Suva when I visited a few weeks later. That was more than a year before the couple stepped down as senior royals and moved to the US.

Fiji's plans for fighting climate change include[8] a goal of net zero greenhouse gas emissions by 2050. It also seeks to restore coastal ecosystems such as mangroves, sea grasses and reefs that can help to offset sea level rise by building up the coasts. Among other measures, Fiji will plant 30 million trees by 2035.

Fiji "lies on the front line of adverse climate change impacts and will be facing some of the most severe climate-related challenges in the coming decade," the plan says. "These range from prolonged droughts, changes in the hydrological cycle resulting in intense floods, and extreme weather events, to rising sea levels and its resultant saltwater intrusion and loss of habitable land," it adds.

"We understand the situation, we are the coal face of it," Fiji's Attorney General Aiyaz Sayed-Khaiyum told me of rising seas driven by man-made greenhouse gas emissions in an interview in a hotel near Nadi, the main international airport, when I visited in December 2018.

Fiji has more than 300 islands,[9] over 100 of which are permanently inhabited, and a population of almost 900,000. It takes about four hours to fly to the islands from Sydney, Australia. Overall, Fiji is less vulnerable than many Pacific states because its main islands are formed by extinct volcanoes, making them higher and more robust than the coral atolls that make up low-lying island nations such as Tuvalu.

"We are perhaps luckier overall than the other atoll countries where it is existential for them, like it is for some of our low-lying communities," Sayed-Khaiyum reflects. He noted that Fiji's highest peak is more than 1,300 metres – Tuvalu's highest point is about 4 metres. Fiji has offered to do its bit to take in other island peoples if rising seas make their homes uninhabitable elsewhere in the Pacific.

Daniel Lund, a special adviser to the Fijian government about climate change, tells me via email that Fiji had partially relocated two other communities before Vunidogoloa after flood damage along the shoreline was caused by blasts from tropical cyclones: seven households were relocated in Nagasauva Village in 2011, and nineteen in Denimanu Village in 2013.

Lund explains that, by contrast, "Vunidogoloa was relocated with state support due to impacts which were clearly driven by slow-onset sea level rise. Efforts at adaptation had been made but eventually it was clear that relocation was the most effective and practical solution. As a result – this village relocation has featured as a sort of 'classical example' of climate change induced relocation." He goes on to say that Fiji needs "standard operating procedures" to guide relocations but that these cannot be a one-size-fits-all. Relocations have to be flexible, and incorporate the wishes of local communities. Fiji is working to identify a shortlist of fifteen of the communities most in need of relocation.

But the pandemic looms over all plans for coming years, undercutting tourism and economic growth. "Due to the COVID-19 crisis, Fiji's economy is expected to shrink by approximately 20 per cent in 2020, with the tourism sector facing the full brunt of the pandemic-induced travel restrictions. As a result, around 118,500 people are jobless," the government states in its climate report[10] to the UN.

Before the pandemic, the number of tourists doubled from 1995 to more than 800,000 a year, almost matching the population. And yet the costs of climate change continue to rise.

Fiji's Climate Vulnerability Assessment (CVA)[11] projected that by 2050, over 6.5 per cent of Fiji's annual Gross Domestic Product could be lost due to tropical cyclones and floods, and that climate change and natural disasters will drive many more people into poverty. Plus, extreme floods that usually hit the city of Lautoka every 100 years, caused by a combination of high tides and storms, could be expected every second year in 2100 if global emissions keep rising.

Despite the rising costs, Fiji has a volunteer army of activists, including young people, to help fight the impacts of climate change. Teenage activist AnnMary Raduva (@annmary_raduva on Twitter) has campaigned against plastic waste, including party balloons being released into the sky, and has worked to plant mangroves with other young people around Fiji's coast to protect it from erosion and rising sea levels. "Satisfying seeing all the grown mussels & an increase in fish stock around the small mangrove flat!" she wrote in one 2021 tweet. She spoke at a UN climate summit in 2019 and was also among a group who met New Zealand Prime Minister Jacinda Ardern at a 2020 meeting in Suva.

Fiji, when it hosted UN climate talks in 2017, did its best to spur action after former US President Donald Trump had announced he would pull out of the deal – a decision reversed by President Joe Biden on his first day in office in 2021.

The lack of hotels in Fiji's capital Suva, which has a population of about 90,000, meant the talks had to be relocated to Bonn, Germany. 'Bula' (pronounced 'boo-lah'), the Fijian greeting for 'hello' that literally means 'life', became widely adopted by delegates at the 'Fiji-on-the-Rhine' conference.

"You probably think this place is tiny," a taxi driver told me when I visited Suva. "But people coming from other Pacific nations for the first time think it's a vast city."

David Boyd, the UN Special Rapporteur on Human Rights and the Environment, urged nations responsible for greenhouse gas emissions to give more international aid to help relocations in Fiji. "Relocating dozens of communities is an expensive process whose costs should not be borne by Fiji but rather by those who caused the problem," he wrote in a 2019 report[12] after a visit to Fiji, including a trip to Vunidogoloa. He is adamant that rich nations should honour a commitment made under international climate agreements to mobilise at least US $100 billion a year for developing nations by 2020 to help protect against climate change.

"Fiji is undoubtedly a global leader in advocating international policies to counter climate change and has developed an impressive array of plans, strategies and policies," Boyd wrote. Yet for all Fiji's good intentions, he said there was a lot of work still to be done, noting that Fiji ranks low – 134th of 180 nations – on a 2020 Environmental Performance Index[13] compiled by Yale University. The index – topped by Denmark, Luxembourg, Switzerland and Britain – seeks to gauge how well countries are protecting the environment around the world, from biodiversity to air pollution. Poorer nations like Fiji, with less money to spend, tend to get lower marks.

Boyd stated that Fiji already devoted about 10 per cent of its budget to reducing climate risks and preparing for natural disasters. Proposed actions to limit emissions could cost US $3 billion, while plans to adapt to climate change, such as moving villages, have an estimated cost of US $5 billion.

Boyd was generally positive about Vunidogoloa and what he said had been a "deeply painful" process of relocation and abandoning the graves of the villagers' ancestors. "The new village site offers improved housing conditions, better sanitation, solar-powered electricity (although currently limited to one small panel per home), and improved access to health care and education. The new site was selected by the community and subjected to an environmental impact assessment," he wrote. Among shortfalls, he noted that kitchens were not installed indoors and that people were cooking on open fires – a health hazard due to the air pollution.

Other researchers have also given the project high marks. 'From Vunidogoloa to Kenani: an insight into successful relocation'[14] was the title of a study by Clothilde Tronquet, a researcher at the Institute for Climate Economics in France. 'A far cry from the negatively connoted representations of environmental migration as being emergency-driven and compelled by humanitarian concerns, the Vunidogoloa relocation is the result of a well-thought-out, multilateral and participative process that lasted nearly a decade,' she wrote. She estimated the Fijian government probably paid about three-quarters of the cost of the village's relocation, with the villagers contributing the rest.

When I asked Botu about such findings, he disagreed, telling me the locals had contributed a bigger share via labour, timber and land.

Tronquet also wrote in her review that the high cost for the government, estimated at almost a million Fijian dollars, "casts doubt on the viability of the project to be replicated elsewhere".

Paradoxically, after big public investments and favourable outside reviews, 90 per cent of people in Vunidogoloa voted for opposition parties in November 2018 elections when Prime Minister Bainimarama won a new term.[15]

Botu says the reason for this is that the government has been too slow to finish the project, leaving the villagers to clean up. Some drains he showed me were incomplete years after an October 2014 deadline set by the government, water pipes lie unburied and houses lacked promised extensions for kitchens, he said. "The government told us 'you have to do it on your own'. So, when the election came we decided we don't vote for them," he explained.

Attorney General Sayed-Khaiyum, however, insists Fiji has kept its side of the bargain. He said many factors were at play in Bainimarama's wafer-thin re-election in 2018 and Vunidogoloa was not untypical in voting for the opposition.

And the recriminations date back a while.

During the 2018 election campaign, the *Fiji Times* newspaper stated that Bainimarama had branded the people of Vunidogoloa ungrateful '*liumuri*' (backstabbers) after they failed to vote for him in the previous election in 2014, despite the splurge of public investment.[16]

Botu pulls out a book signed by Bainimarama during a visit to Vunidogoloa in January 2014, when the village opened, and frowns at the memory. "They didn't do everything they promised."

People are rightly reluctant to move because of factors such as climate change that are beyond their control. Moving from ancestral lands is a traumatic process even when the move is likened to a biblical journey to a promised land.

Illustrating that point, Jane McAdam, a law professor at the University of New South Wales in Australia and a leading expert on migration, wrote in a 2015 review[17] that people uprooted from their homes are generally unhappy about the dislocation: "The history of relocation is characterised by a gulf between grand theoretical visions on the one hand and the challenges of practical implementation on the other." She observed, "Resettlement is a fraught and complex undertaking, and rarely considered successful by those who move."

Another study[18] led by Annah Piggott-McKellar of the University of Queensland found that women felt marginalised in the move of Vunidogoloa and the other Fijian community, Denimanu, where nineteen families were moved. She quoted one woman in Vunidogoloa as saying – anonymously – "For us, the women, we just listen to whatever the men say and we just agree. They never consult us. The voice of the men is the only voice that is heard, so we just listen to that voice. So whatever the men has agreed we just consent to it."

In Fiji's Planned Relocation Guidelines,[19] Prime Minister Bainimarama also says that climate change is worsening day by day. "The effects of climate change are clear to most Fijians. Rising of seas continue to erode shorelines and encroach on coastal communities, and Tropical Cyclone Winston tore through Fiji with unprecedented strength in 2016, causing damages amounting to one-third of Fiji's Gross Domestic Product," he wrote in the guidelines. Forty-four people died in the storm he referred to. Meanwhile, Fiji's guidelines say that relocation is "an option of last resort" to be considered "**ONLY** when all adaptation options have been exhausted". (The guidelines put **ONLY** in upper case and bold). It adds, "It is expected to become a more common response to climate related events in the future."

Fiji has gone a long way to working out that relocation is much more than just moving people away from a vulnerable region. It also has to ensure longer-term livelihoods. The Fijian guidelines formally define relocation as "the voluntary, planned and coordinated movement of climate-displaced persons within States to suitable locations, away from risk-prone areas, where they can enjoy the full spectrum of rights including housing, land and property rights and all other livelihood and related rights. It includes: displacement, evacuation (emergency relocation) and planned relocation."

In some ways, Fiji's vulnerability is linked to a history of colonial settlement, which favoured encouraging people to live by the coast rather than inland, and this is now being aggravated by tourism. Previously, many Fijians traditionally lived inland where they were less vulnerable to seaborne attacks by rivals. In the Pacific, for example, "while traditional settlements on high islands [...] were often located inland, the move to coastal locations was encouraged by colonial and religious authorities and more recently through the development of tourism", the IPCC said in a 2019 report.[20] Such population movements are small but "play a critical role at the very local scale" in explaining why some people are more exposed to the risk of sea level rise.

Despite Prince Harry's assertion that Vunidogoloa was the first village moved inland due to sea level rise, there are sadly others that could contest that title. In 2005, the UN Environment Programme said[21] a community of about 100 people on Tegua Island in Vanuatu, north-west of Fiji, "has become one of, if not the first, to be formally moved out of harm's way as a result of climate change" by moving to higher ground. There, the islanders moved about 600 metres inland to avoid coastal floods.

Ursula Rakova, leader of Tulele Peisa ('sailing the waves on our own') group which helps with relocation from the Carteret Islands[22] in Papua New Guinea, says climate change is making the atolls uninhabitable. "We can no longer grow the food crops we used to grow on the island some 40 years back. There was an intrusion of salt water into the land. Leaves were turning yellow," she states in a video[23] published by The Conversation in 2019. "Our fruit trees were also beginning to die away. And then it results in a lot of women and children also getting sick, with dysentery, malnutrition." She explains, "We started to move, to relocate people in 2009, and we are continuing to do that. We have a population of 2,700 people, meaning we will have to move 150–200 families over to mainland Bougainville."

The World Economic Forum, in 2017, wrote[24] that rising seas were threatening many parts of the world "but for some communities, it's already too late to save their homes". It went on to list "five places with a plan B". Three are in the Pacific – Vunidogoloa, Kiribati and Taro in the Solomon Islands – and two in the US – Shishmaref in Alaska and Isle de Jean Charles in Louisiana, where wetlands are sinking into the Gulf of Mexico.

Meanwhile, Bainimarama warned the UN in a speech in Bonn that rich countries should not reckon that remote places on the other side of the world were the only ones at risk from rising seas. "Cities in the developed world like Miami, New York, Venice or Rotterdam" are as vulnerable to the global rise in sea levels as low-lying nations, he warned.[25] This was a veiled jab at Trump, who owns properties in both New York and Florida and doubts that climate change is man-made.

Fiji is part of an international coalition[26] led by small island states, that wants deep cuts in greenhouse gas emissions to limit global warming to 1.5°C above pre-industrial times as part of the global 2015 Paris Climate Agreement. Average world temperatures are already up about 1.2°C and on track to breach the limit in coming years.

Fiji also foresees the complexity of planning to protect not just people but infrastructure on the coast. "Maritime transport has been described as the lifeline of Pacific small island developing states, such as Fiji," the country states in its 2050 climate plan. "By definition, all maritime infrastructure (which includes: ports, jetties, access roads, navigational markers and beacons, warehousing, etc.) sit at or very close to sea level, and will be the first and most affected by rising seas, king tides, and storm surges," it warns.

In 2013, Fiji also adopted a progressive constitution[27] to protect the environment. Article 40 of it reads: "Every person has the right to a clean and healthy environment, which includes the right to have the natural world protected for the benefit of present and future generations through legislative and other measures."

Boyd believes that Fiji's new constitution could be an example to other Pacific states. "The inclusion of the right to a healthy environment in the Constitution of Fiji sets an important precedent for other small island developing states, as many of these nations have not yet granted legal recognition to this fundamentally important human right," he wrote in his 2019 report.

Fiji's government says it was also the first to promise to take in potential Pacific migrants from lower-lying atolls – such as Tuvalu or Kiribati – if entire islands become uninhabitable this century. Certainly, Fiji has more space than many other Pacific states. It's about the size of Israel or the US state of New Jersey, with a land area of 18,000 sq km. That means there are 50 people per square kilometre on Fiji, compared to a far greater population density of more than 1,000 people per square kilometre in Tuvalu, or about 400 in Kiribati.

Among initiatives to protect livelihoods, Kiribati, a nation of thirty-three Pacific islands, bought 6,000 acres (2,400 hectares) of forest land in Fiji in 2014 for A$9.3 million (US $8.3 million) on the island of Vanua Levu, west of Vunidogoloa. The plan is to use the land for farming to produce food, and as a possible refuge if seas rise in coming decades and make some of Kiribati's atolls uninhabitable. "I am not going to relocate my people, but someone else might in coming decades," former Kiribati President Anote Tong told me in an interview[28] for Reuters at the time.

Sea level rise is hugely complex – cyclones and shifting currents can wash sediments around, even sometimes forming new or bigger islands despite the creeping rise in ocean levels. In healthy oceans, corals can also grow upwards and keep up with sea level rise. Mangrove trees and shrubs can also keep pace with rising seas – up to a point. A 2020 study,[29] in seventy-eight coastal sites from the Ganges to the Gulf of Mexico, showed that mangroves thrived at the end of the last Ice Age only when rates of sea level rise were less than about 7mm a year. Global rates of sea level rise are now about 3.7mm a year and could exceed 15mm a year with high rates of growth in greenhouse gas emissions by 2100. The 7mm "threshold is likely to be surpassed on tropical coastlines within 30 years under high-emissions scenarios," a group of scientists, led by Neil Saintilan at Macquarie University, New South Wales, Australia, wrote in the journal *Science*.

However, studies about islands losing or gaining ground are often hugely controversial. Powerful storms or shifting ocean currents can allow sediments to build up some low-lying islands, apparently in defiance of the global rise in ocean levels, even as others vanish beneath the waves. One 2018 report[30] showed that the overall area of the Pacific island nation of Tuvalu, with 10,600 people living on 101 islands, had a net gain in land of 73.5 hectares over the past four decades, a 2.9 per cent increase. That report, in *Nature Communications*, was wrongly seized upon by climate sceptics as evidence that the entire idea about 'sinking islands' was fraudulent. Among them, late US radio host Rush Limbaugh said, in a radio segment headlined 'More Evidence Climate Change is Fake':[31]

> Do you remember the global warming stories of years ago ... There was this little island in the South Pacific, and it was shrinking, and they went out there and they had this mayor standing on ... It was like a polar bear on a shrinking iceberg. They had this mayor standing on what was left of the town hall or what. Global warming, rising sea levels were about to swallow up this poor little guy's island nation. And it was used to show that climate change is happening and this is just a precursor to what's gonna happen to New York. There's a problem. This island nation is called Tuvalu, and all of a sudden it's growing. It's expanding. It is getting bigger. It's popping up out of the ocean. It is getting back to where it was. And this is confounding the climate change crowd. It doesn't make sense to them. This should not be happening.

Tuvalu's then Prime Minister Enele Sopoaga hit back,[32] saying the study should have consulted Tuvalu and included effects of climate change, such as ocean acidification that may be undermining reefs or coral bleachings caused by rising temperatures. "I find it totally unfortunate and perhaps untimely in that the news item was never allowed responding or verifying by Tuvalu authorities," he said. "It was simply put for public consumption without proper verification. I strongly feel this is irresponsible reporting and irresponsible disclosure of information which should have been properly verified."

The study itself was more nuanced. It began: "Sea-level rise and climatic change threaten the existence of atoll nations. Inundation and erosion are expected to render islands uninhabitable over the next century, forcing human migration." Still, it also said the results of the study "challenge perceptions of island loss, showing islands are dynamic features that will persist as sites for habitation over the next century". Indeed, lead author Paul S. Kench of the University of Auckland in New Zealand told the *New York Times*[33] that the impacts of climate change were complicated: "Our work isn't suggesting they have nothing to worry about ... But there is more to this than the simple, linear doomsday scenario."

But the acceleration of sea level rise may jeopardise islands.

"There is growing concern that some island nations as a whole may become uninhabitable due to rising sea levels and climate change, with implications for relocation, sovereignty and statehood," the IPCC concluded in a 2019 overview of the risks.[34]

It noted that many low-lying islands shift – a study of 709 islands in the Pacific and Indian Oceans found that 73.1 per cent had stable surface area while 15.5 per cent had gained and 11.4 per cent shrunk over the last forty to seventy years. But it cautioned that the trend is set to shift to more disappearances as seas rise. A quickening pace of sea level rise, changes in storms and waves, ocean warming and acidification of reefs "is expected to shift the balance towards more frequent flooding and increased erosion".

"Make no mistake [...] climate change's existential threat to our own survival is not a future consideration, but a current reality," Gaston Browne, Prime Minister of Antigua and Barbuda and chair of the Alliance of Small Island States (AOSIS), told the UN Security Council[35] in early 2021. "For the past 30 years, small island and low-lying states have been sounding the alarm, sending the S.O.S. distress signal. We are losing our territories, populations, resources, and very existence due to climate change. The world is also facing this threat, and it is not by weapons, but by an enemy we all agree is real – climate change," he said.

Back on the beach in Fiji, Botu sees no hope of halting the rise of the ocean that has already forced his village to move from ancestral land on the coast. He simply reflects that, "Climate change cannot be stopped, it will be increasing, increasing."

3

AN ANDEAN GLACIER IS MELTING: SEE YOU IN COURT

Seas are rising. So are we.

Youth activist banner in climate lawsuit

A Peruvian farmer is suing German power utility RWE, saying its greenhouse gas emissions are contributing to melting an Andean glacier that could bury his home with a cataclysmic mudflow. RWE rejects the charges, but the lawsuit raises a question: who's to blame for climate change? Lawsuits are spreading worldwide – in the US, young people say rising seas violate their constitutional rights. Can courts solve climate change?

HIGH IN THE PERUVIAN ANDES ten black plastic pipes fan out into an icy blue lake, sucking away water to limit risks of a flood from thawing glaciers that blanket the mountains above. This is Lake Palcacocha, one of the most menacing places in South America. At a lung-bursting altitude of 4,560 metres, Palcacocha lies in the shadow of dazzling glaciers, crumpled and fissured rivers of ice clinging to some of the highest peaks in the Andes.

In 1941, an outburst flood from Palcacocha swept down the valley and dumped mud, ice and rocks onto the city of Huaraz, 23km away. At least 1,800 people died,[1] although some estimates put the toll much higher, at perhaps 7,000 people.[2]

Dangerously swollen again, the lake is the focus of a landmark lawsuit in Germany brought by a Peruvian farmer who says industrial firms on the other side of the world can be liable for global warming, and the melt of glaciers like those hemming in Palcacocha. He wants money to build better protective walls.

Workers monitoring the 1.5km-long lake, as part of an early warning system, say they often spot condors soaring above, mesmerising black dots in the chill air. A few hardy yellow flowers, low shrubs and grasses grow near the lake, but much of this remote Andean valley is a barren wilderness. The ten pipes – each about 25cm in diameter – are jarringly out of place, like a stranded giant octopus with two extra tentacles extending into the milky blue waters.

Scientists say glaciers here are melting faster than elsewhere, as the effects of global warming creep higher into the last tropical mountain strongholds of ice in Latin America. The extra pulse of meltwater is already disrupting farming and hydropower generation down the valleys.[3]

The ten plastic tubes dangle in the icy water well, tethered to floating blue barrels, as they siphon glacier meltwater from the tear-shaped lake and carry it up and through a hole in a low dam, with a gurgling and rumbling noise caused by trapped air bubbles. The pipes plunge a couple of hundred metres down the mountainside and around a corner where they end, spewing out a foaming torrent bound for Huaraz and beyond to the distant Pacific Ocean.

Trying to prevent a repeat of the 1941 disaster, Saúl Luciano Lliuya, a farmer, mountain guide and father of two children, sued German utility RWE AG in 2015, saying RWE should help pay to upgrade the flood defences and protect his family and home in a danger zone on the outskirts of Huaraz. His logic is that RWE is historically Europe's biggest greenhouse gas emitter and so its coal-fired power plants are partly to blame for the glacier melt on the other side of the globe.

The lawsuit estimates that RWE has emitted about 0.5 per cent of all global greenhouse gases since the Industrial Revolution and so the company should pay €17,000 towards the upgraded defences, or 0.5 per cent of the total estimated cost of US$3.5 million.

RWE, founded in 1898 as the Rheinisch-Westfälisches Elektrizitätswerk, says the case "has no legal basis".[4] It also says it's impossible to join the dots from burning coal in Germany to the thaw of a glacier half a world away.

A district court in Essen, Germany, accepted RWE's argument and dismissed the case in 2016. But, after Luciano Lliuya appealed, the Higher Regional Court of Hamm ruled in 2017 that experts should collect evidence in the case. With multiple delays and coronavirus, they haven't yet done so. It could take years to resolve.

Still, "This lawsuit is the first of its kind in the whole world to reach this stage," according to Germanwatch, an independent development and environmental organisation based in Bonn, Germany, which backs Luciano Lliuya.[5]

Certainly, companies worried about the risks of liability for climate change are closely watching the case, even though it's a legal long-shot. A ruling in Luciano Lliuya's favour could drive big emitters to bankruptcy if it led to a slew of successful copycat suits. It would make billion-dollar payouts by tobacco companies, for the harm caused by smoking, look like small change.

I meet Luciano Lliuya one evening for dinner in a restaurant and bar in Huaraz in late 2018. He's a fit, wiry man who hopes the case is important far beyond Palcacocha as a lever for environmental change in the world.

"As people, as families, we want a clean environment. It's our right to have a future with a good climate," he says. "The problem is that the economic system is based on dirty energies. To change this system of dirty energies we have to raise pressure. That's what we're doing with the lawsuit – putting pressure on politicians to change the dirty economy." His concern is that the current defences for Huaraz, such as the octopus siphon, are simply not up to the task of staunching vast glaciers.

One of Luciano Lliuya's friends, mountain guide William Beltrán, offers to accompany me on the 1,500-metre climb from Huaraz to Palcacocha, and one of his contacts, Hugo Miranda Quinones, has a robust four-wheel drive car to take us up on the snaking, often precipitous dirt road. We set off early one morning, past the suburbs and adobe homes on the edge of the city, chased by dogs multitasking as they try to nip the tyres while running, barking furiously and wagging their tails. The gears grind up past groves of eucalyptus and pine trees, past fields for potatoes, wheat or quinoa, over rickety wooden bridges into a narrowing gorge. Huge boulders by the roadside force sharp, heart-stopping bends almost over the void. A large deer runs away startled in the distance, a few cows stand munching grass in a valley where a pack of wild dogs roams. It's ever uphill, the car straining all the way.

It requires a permit from the Huaraz city authorities to drive up here, past a padlocked metal gate. An official living in a house nearby produces a key and lets us through, after we show her our permit. Locals grimly say that this Cordillera Blanca, part of the biggest set of tropical glaciers in the world down the spine of the Andes, will soon have to be renamed part of the Cordillera Negra – the darker, glacier-free range of mountains in the distance away to the west.

It's hard to imagine how any mass of water or mud from Lake Palcacocha could manage to plunge all the way to the city. The distance seems too great, the valley too wide with momentum-sapping flatter stretches. But it did, on 13 December 1941,[6] when, early in the morning, a glacier overhanging the lake cracked and triggered a landslide that sent a flood of water down the valley, in turn bursting another lake on its way, Jiracocha, which combined forces to swamp Huaraz.

After reaching the end of the road, it's a faint-inducing couple of hundred metres walk up to Palcacocha. As I gasp, Beltrán skips up as if the oxygen is as thick as at sea level – he's grown up here and has conquered all the Andes' highest peaks, including Aconcagua in Argentina, the world's highest mountain outside Asia at 6,962 metres. After seeing no one in the padlocked valley for 10km, we meet Juan Morales at the rim of the lake, a 53-year-old whose face is wrinkled and darkened, he doesn't use sunglasses despite the blazing Andean sun.

Morales is one of a three-man team monitoring Palcacocha, twenty-four hours a day, seven days a week, all year round. Their job, living here for two weeks at a time, is to report every two hours on the temperatures, winds, cracks in the ice and any chunks falling off the mountainside – a human early warning system for their families and thousands of others living below. The night shift in the Andes surrounded by the creaks and groans of glaciers sounds a lonely place but Morales says, "I'm not scared to be working here. We're protecting the city."

At Palcacocha, it's so high that the lack of oxygen means your brain doesn't work as well as it does lower down. At these altitudes, scientists say you'd get a better view of the stars sparkling at night from lower down the mountainside because your eyes and brain function better with more oxygen in your blood.

A mobile phone transmitter has recently been set up on the local mountainside. It will enable the guards to call emergency services to set off evacuation sirens in Huaraz about fifteen minutes before any mudslide reached the town. Morales says every second counts.

The three guards live together in a stone hut perched about 30 metres above the end of the lake – high enough, in theory, to be out of harm's way in case a tsunami is triggered by a glacier crashing into the lake. "A lot of ice fell into the lake last night," Morales says, pointing to the end of the lake that has grown to 1.5km long from less than 400 metres in the early 1970s. Luciano Lliuya's original lawsuit[7] says that the volume of the lake swelled thirty-fold to 17.3 million cubic metres of water in 2009, from about 580,000 cubic metres in 1972. In the same period, the surface of the lake expanded eight-fold.

Still, Morales says that the octopus siphon pipes installed since 2009 are helping to keep water levels down. To prove it, he points to a barren band of ground just above the water line where bushes and other vegetation haven't yet grown since the water receded in recent years. "We're keeping the water level down to protect the city," he reiterates. When the lake burst in 1941, it carried with it more than 10 million cubic metres of water, rocks and ice.[8] Now, Morales reckons that the siphon has drained about 3 million cubic metres of water from Palcacocha – equivalent to 1,200 Olympic swimming pools – reducing it to 14 million cubic metres in late 2018 in just under ten years. Among other evidence of increased monitoring, a solar panel was installed in 2018 to help run an automatic gauge in the lake that will react to a sudden change in the water level.

Morales says the safest spot to take photographs is a small knoll to the north side of the lake, but don't venture down along the southern side, he cautions – there are rockfalls all the time from the steep cliffs.

Palcacocha, lying below the mountain peaks of Nevado Palcaraju (6,274 metres) and Nevado Pucaranra (6,156 metres) that seem to pierce the sky, is just one of many high-altitude lakes around the world where eternal snows are falling short of their name. And the thaw is adding to the risk of glacial lake outburst floods (GLOFs). Receding glaciers in mountain ranges such as the Andes, the Alps, the Rockies or the Himalayas are all releasing extra meltwater that is often trapped in high-altitude lakes like Palcacocha. Peru itself has more than 75 per cent of the world's tropical glaciers,[9] which would contribute a tiny 0.2mm sea level rise if they all melted.

Glacier lake outburst floods are often deadly, the IPCC said in a 2019 report:[10] "Glacier lake outburst floods alone have over the past two centuries directly caused at least 400 deaths in Europe, 5,745 deaths in South America, and 6,300 deaths in Asia," it says, adding that individual disasters such as in Huaraz dominate the death tolls.

At Palcacocha, an earthquake could destabilise steep boulder-strewn slopes around the lake, causing a landslide and a GLOF that sweeps over or destroys the dam. Or the glaciers above could unleash a flood if a large section collapses. Up close, the brownish grey slopes surrounding the lake look like a loose pile of mud and rocks, somehow glued together. It's hard to imagine the danger in this remote spot, silent except for when a worker yanks a diesel-powered portable generator into life to help ensure the siphons keep flowing.

Under the terms of the 2017 ruling[11] by the Higher Regional Court of Hamm, experts will have to come here and, among other things, they will have to find out whether "a flood and/or mudslide resulting from the significant expansion and increase in the volume of water in Lake Palcacocha poses a serious threat" to Luciano Lliuya's property. The court also requested scientific opinions about whether emissions in Germany could be linked to the melt.

Meanwhile, RWE insists that it cannot be held liable for the problems of melting ice in Peru. "Mr Lliuya's claims against our company have no legal basis and do not comply with German civil law," Guido Steffen of the RWE press office wrote in an email when I asked for the company's views in 2021. "Climate is an extremely complex matter and shows an abundant natural variety and diversity. Therefore, to our opinion, it judicially is impossible to relate specific/individual consequences of climate change to a single person responsible," he said. He added that governments and treaties, such as the 2015 Paris Agreement, should provide solutions, rather than the courts. And he said RWE was abiding by local German laws regulating emissions. He added that RWE "supports the German and European Union climate goals and agrees to

Germany's coal phase-out that will cease the usage of coal by 2038." Also, he said RWE will work to be carbon dioxide neutral by 2040 by cutting emissions and RWE was investing €5 billion in more windfarms, solar parks and other green measures by the end of 2022.

In court documents, RWE also says that glaciers in the Andes have grown and retreated unpredictably for centuries amid natural swings in the climate. Among other arguments, RWE notes that the overflow siphon pipes have helped to reduce the water levels in recent years and the company "contests that there is an acute flood risk". Drawing on precedents abroad, in its defence, RWE likened the case to an unsuccessful 2008 attempt by Kivalina,[12] a native village in Alaska, to sue ExxonMobil Corp and other energy companies for global warming that, they said, was contributing to floods and erosion. US courts dismissed the Kivalina arguments. RWE also points to other factors that could contribute to local warming around Palcacocha. Huaraz city has grown to a bustling 120,000 population now, from about 20,000 in 1941, and the air is thick with pollution. Maybe dark soot is settling on the glaciers around Palcacocha, blackening the ice and making it more vulnerable to melt. RWE also points out that a ruling for Lliuya would open a legal free-for-all of claims and counter-claims. "If it were possible to ascertain liability in this manner, every individual person could be held accountable as a disturber and would qualify as an injured party at the same time. The result would be a liability of 'all against all', which exceeds the regulatory limits of German civil law," it says in one legal filing.

In the first case, the Essen court ruled that the claim was "not sufficiently precise", according to a translation by Germanwatch.[13] It largely echoed RWE's defence that it was impossible to link emissions in Germany to the swelling of an Andean lake. "The pollutants, which are emitted by the defendant, are merely a fraction of innumerable other pollutants, which a multitude of major and minor emitters are emitting and have emitted. Every living person is, to some extent, an emitter," it said. "When innumerable major and minor emitters release greenhouse gases, which merge indistinguishably with each other, alter each other, and finally, through highly complex natural processes, induce a change in the climate, it is impossible to identify anything resembling a linear chain of causation from one particular source of emission to one particular damage," it added. The court also reasoned that Lake Palcacocha would probably still be dangerously swollen if RWE had never existed. And it said that the measures outlined by Luciano Lliuya might not be sufficient to prevent any future floods.

Down in Huaraz when I meet Luciano Lliuya, he's encouraged by the Hamm court ruling that requires experts to visit, but says a lot needs to happen to protect the city. In the bar, the television is showing Peru's soccer team playing the US with the Peruvian commentators loudly praising every successful Peruvian pass. It ends 1–1 with loud cheers after Peru claw their way back with a late equaliser by Edison Flores. Looking around the fans in the bar, Luciano Lliuya says there's a danger that many locals are growing complacent about Palcacocha, forgetting about what happened in Huaraz in 1941. "My father told me when I was a child about the mudslide. He was one year old when it happened and heard about it from his parents. I learnt about it in the same way. As generations go by it gets forgotten. History gets forgotten. We have to remember," he says.

Even he acknowledges that his case is a struggle because there is scant legislation about harm caused by greenhouse gas emissions. "A judge needs to be able to say 'you've broken the law'," he says with a shrug. When there are no relevant laws, a judge has limited options even if she or he is sympathetic to the plaintiff.

Still, Luciano Lliuya reckons that attention to the case around the world means that it is already a victory, no matter the outcome. And he says any award by a German court will go to local authorities to build flood defences. One problem in Huaraz, he says, has been that some locals thought he was jetting off to Germany for court hearings and stood to pocket €17,000 if he won. "The criticism was that I was personally going to gain, that I was going to Germany to sell the lake, to sell the water. The people didn't understand how this lawsuit could benefit to reduce temperatures," he said. Things are now changing, he says, and locals are more enthusiastic about supporting his case.

He won a boost in February 2021 when a report in the journal *Nature Geoscience* linked regional warming in the Cordillera Blanca around Palcacocha to rising temperatures, even as long ago as when the 1941 disaster struck in Huaraz. It was at least 99 per cent certain, the scientists wrote, that the retreat of Palcaraju glacier above Palcacocha by 1941 "represented an early impact of anthropogenic (man-made) greenhouse gas emissions". The main estimate was that the overall retreat of the glacier "is entirely attributable to the observed temperature trend" rather than to natural variations in the climate, according to scientists at Oxford University and the University of Washington in Seattle. Average surface temperatures in the region rose by about 0.25ºC by 1941, compared to levels in the late nineteenth century, and by 1.3ºC by 2019. That warming trend melted glacier ice at slightly higher altitudes every year, they wrote. The extra water then flowed into the lake, slowly adding to the risks.[14]

"Climate change has played a role. It is implausible that the glacier would not have been responding to the warming," lead author Rupert Stuart-Smith, of the Environmental Change Institute at Oxford University, told me on a Zoom call. He added that greater monitoring was needed around the glaciers of the world, to alert to the risks, such as a flood in the Himalayas in India in February 2021, that left more than 200 people dead or missing.[15] Satellite monitoring might see cracks appear in glaciers or on mountainsides at risk of landslides that can spill into rivers.

In places such as Palcacocha, Stuart-Smith said that glacier melt is the key underlying risk even if the trigger of the outburst flood has nothing to do with higher temperatures, such as an earthquake. "Climate change is the cause of the risk," he said. Despite RWE's arguments that soot might be blackening the glacier, other ice far from cities in Peru was shrinking at a similar rate.

"Around the world, the retreat of mountain glaciers is one of the clearest indicators of climate change," added Professor Gerard Roe, of the University of Washington, who was also an author of the study. "Outburst floods threaten communities in many mountainous regions, but this risk is particularly severe in Huaraz, as well as elsewhere in the Andes and in countries such as Nepal and Bhutan, where vulnerable populations live in the path of the potential floodwaters," he wrote in a news release about the findings.

The *Nature Geoscience* study did not, however, take a view on whether blame can be traced back to individual emitters, such as RWE. "This type of science hasn't really had its day in court yet," Stuart-Smith said.

That day may be coming with a growing branch of climate 'attribution science' – linking risks of individual events such as heatwaves and downpours to greenhouse gas emissions. Slow-motion changes, such as increasing glacier melt, the shrinking of Arctic sea ice and rising seas, are more clearly linked to global warming.

For attribution science, researchers say greenhouse gases are like loaded dice for the climate system, raising the risks of extreme weather events beyond what would happen naturally. Yes, a catastrophic GLOF might have happened anyway, without human influences, but climate change has made it more likely. By analogy, you might by a fluke throw several sixes in a row, but your chances are a lot better with weighted dice. Scientists say such attribution science has made a lot of headway this century as models of the weather and climate have improved.

A groundbreaking study was conducted into the 2003 heatwave in Europe,[16] probably the hottest in 500 years in which more than 70,000 people died.[17] It could have been a freak event but scientists led by Peter Stott of the UK Met Office concluded that human greenhouse gas emissions had "at least doubled the risks".

Since then, there have been more than 405 studies of attribution for extreme weather events and trends.[18] At a basic level, scientists have known for more than a century that an increase in greenhouse gases can warm the planet. Without carbon dioxide, methane, water vapour and other natural greenhouse gases, the planet's average surface temperature would be an inhospitable -18°C, rather than the current 15°C or so. But the fast-rising concentrations of these heat-trapping gases are making them harmful, disrupting the climate system.

Carbon dioxide, the main man-made gas, makes up about 419 parts per million (ppm)[19] of the atmosphere at the world's best-known measuring site on a mountaintop in Hawaii in 2021, up about 50 per cent from 280ppm before the Industrial Revolution. Those extra gases are the main driver of rising temperatures.

Around the world, the number of lawsuits about all aspects of climate change is surging, according to a 2021 UN Environment Programme report.[20] It found 1,550 climate change cases in thirty-eight nations in July 2020, almost double the 884 cases brought in twenty-four countries three years earlier. Of the 2020 total, 1,220 were filed in the US, according to the tally, compiled with help from the Sabin Centre for Climate Change Law at Columbia University.

Among landmark climate rulings in recent years, the Dutch Supreme Court in 2019 ordered the government to do far more to cut greenhouse gas emissions to protect the human rights of its citizens from the impacts of global warming. Dutch group Urgenda,[21] which campaigns for a shift to renewable energies, said its case was the "first in the world in which citizens established that their government has a legal duty to prevent dangerous climate change".

On the coast of Florida in the US, Levi Draheim is a plaintiff in a legal action seeking to force the state to shift from fossil fuels. He argues that more powerful storms, rising sea levels and other environmental damage violates his rights under the Constitution.

Wearing a blue T-shirt emblazoned "Black Lives Matter", he tells me on a Zoom call he feels "sadness and anger" that his young sister will inherit a planet where the seas may rise more than a metre this century. It's an emotional jolt – Draheim is aged just 13. And his sister Juniper is a nine-month-old infant when we speak in 2021. "It's really upsetting to me ... I want to be able to tell her that I did everything I can to fight these horrible things that we are inheriting. It just makes me sad that people would do these things and know that children will have to deal with it, that their kids will have to deal with this," he says.

Draheim, whose Twitter profile includes a picture of him protesting beside a piece of cardboard emblazoned "Seas Are Rising. So Are We", says he is working to protect the coast. With his mother, "We plant sea grass because it helps to hold the dunes together so that they won't erode when there's hurricanes. The roots hold the sea grass together. It's really helpful for the dunes."

He is one of eight young people in a youth lawsuit filed in 2018 against the state of Florida, backed by Our Children's Trust, a non-profit public interest law firm which has launched climate lawsuits and legal actions in all fifty US states in the past decade. None have got to trial by 2021, and many have been dismissed. "The complaint asserts that in causing climate change, the state of Florida has violated the youngest generation's constitutional rights to life, liberty, property, and the pursuit of happiness, and has caused harm to Florida's essential public trust resources, such as beaches, coral reefs, marine life," Our Children's Trust says.[22]

Draheim had to evacuate his first home, on a barrier island about a metre above sea level, during storms including Hurricane Irma in 2017 when he was "literally up to his knees in flood water and had to put sandbags around the house to protect it from water damage", the legal complaint[23] states. His family have since moved a few kilometres inland, to the city of Melbourne, to slightly higher ground.

"Hurricanes – it is absolutely horrible to see – the most scary thing you can experience because you don't know what is going to happen," he said. He also said that US President Joe Biden's plan[24] to halve US greenhouse gas emissions by 2030, from 2005 levels announced in 2021, was "not enough. It's progress, though." One of his concerns is that fertiliser for grass, washed off lawns in the area during rains, is causing toxic blooms of algae in the Indian River lagoon where he goes sailing and swimming. "The fertiliser is designed to keep plants alive and it is keeping the algae alive," he said.

In 2020, a judge dismissed[25] the Florida case, ruling that the "claims are inherently political questions" and that it was up to politicians, rather than the courts, to solve them.

In May 2021, the First District Court of Appeals upheld that dismissal. Andrea K. Rodgers, Senior Litigation Attorney at Our Children's Trust, said the young activists will keep working to persuade courts that Florida's fossil-fuel energy system is unconstitutional.

"Another lawsuit is in the works" she told me. "So as one door closes, another one opens." She noted that many scientists say that a sweeping transformation of the economy is needed to reduce the concentrations of carbon dioxide in the atmosphere to 350 parts per million, from the current 419, to help cool the planet.[26] That would require a dramatic shift from fossil fuels, while extracting carbon from the atmosphere. In her opinion, Biden's goal of halving US greenhouse gas emissions by 2030 is inadequate, "Any scientist will tell you it is like cutting out half a tumour and telling a cancer patient they'll be ok," she argues, in a Zoom call with me.

Another case that has seen progress is one in which Our Children's Trust backed twenty-one plaintiffs, all age 19 or younger, in a lawsuit in 2015 in the federal district court for the District of Oregon against the US. Similar to the arguments in Florida, the *Juliana v. United States* lawsuit says the youths' rights to life, liberty and property enshrined by the Fifth Amendment to the US Constitution were at risk from climate change. The case almost came to trial in 2018, despite opposition from the Trump administration and fossil fuel companies. Even the US Supreme Court declined[27] a government request to dismiss the case, although it said the breadth of the claims involved was "striking" and that there were "substantial grounds for difference of opinion". In 2020, a panel threw out the case again, but it is still pending in the district court in Oregon.

Among other initiatives to get courts to protect the environment, Pope Francis[28] wants to make 'ecocide' an international crime. He defines ecocide as "the loss, damage or destruction of the ecosystems of a given territory, so that its utilization by inhabitants has been or can be seen as severely compromised". Ecocide "is a fifth category of crimes against peace, which should be recognised as such by the international community," he told the International Association of Penal Law in a 2019 speech.

Currently, the International Criminal Court (ICC) recognises four core crimes: genocide, crimes against humanity, war crimes and crimes of aggression. There are many hurdles to adding ecocide as a fifth – two-thirds of the countries that are members of the ICC would have to adopt an amendment to the Rome Statute that governs the ICC.

Still, the idea of punishing ecocide is gaining ever more attention. In 2021, France's National Assembly voted to make domestic ecocide an offence[29] with penalties of up to ten years in jail, stopping short of environmentalists' calls to define it as an international crime. Celebrities including songwriter Paul McCartney and Swedish climate activist Thunberg back a Stop Ecocide campaign to make environmental destruction an international crime.

Elsewhere, some courts are becoming more active in environmental cases.

In 2021, the German Constitutional Court ruled[30] that the German government should take tougher action against climate change. And a Dutch court told[31] oil company Shell to cut its carbon dioxide emissions from 2019 levels by 45 per cent by 2030, far deeper than the firm planned.

For Peruvian farmer Luciano Lliuya, lawyers say it makes more sense to file a suit against a company in Germany, where most people accept the science of climate change, than to attack a big

US company since there the issue is mixed up in politics. Supreme Court Justice Amy Coney Barrett, for instance, said during her confirmation hearing in October 2020 that climate change is "a very contentious matter of public debate". She added, "I will not express a view on a matter of public policy, especially one that is politically controversial, because that is inconsistent with the judicial role."

Nevertheless, Luciano Lliuya's case has wide international attention. "The sum in dispute may be less than €20,000, but the precedent-setting potential is clear," international law firm Freshfields Bruckhaus Deringer wrote of the Peruvian case in an overview[32] of climate legislation around the world.

In response to RWE's objection that a ruling in favour of Luciano Lliuya could open the floodgates to similar cases, Noah Walker-Crawford, an advisor to Germanwatch on climate legislation and who has written a doctoral thesis about the Peruvian case at Manchester University, said to me in a telephone interview that the goal was not to bring down major corporations but to foster change. "The solution is not for every small-scale farmer to sue every big energy company. We need solutions at a political level," he said. "The only reasons that litigation is going on is because there is a lack of political solutions for people like Luciano Lliuya." Plus, it would be too complex and costly for similar plaintiffs to try to sue. Walker-Crawford was one of the originators of the legal action, travelling to Huaraz after a UN climate meeting in Lima in late 2014. "Litigation helps create pressure to push industry towards more sustainable forms of energy production," he explains. "A few years ago it looked like proving this causal link between RWE and the flood risk in Peru could be quite difficult. On the scientific side it's really turned around since then."

In a document saying, "Glaciers are melting. Responsibility is growing", Germanwatch lists its goals[33] as:

1) To hold polluters like RWE accountable and incite them to shift to less damaging business models.

2) To support the claimant and citizens of Huaraz in reducing the risk of a disastrous flood.
3) To bring about national and international political solutions for protecting those who are most vulnerable to climate change.
4) To develop new legal mechanisms for people affected by climate change – as leverage for political solutions.

Luciano Lliuya's claim that RWE is responsible for 0.5 per cent of historical emissions is based on studies by the US based Climate Accountability Institute, run by American scientist Richard Heede. Many companies around the world emit far more greenhouse gases than RWE. According to a December 2020 report by the institute,[34] twenty corporate emitters accounted for 35 per cent of all carbon dioxide and methane emitted from 1965 to 2018, primarily due to the burning of their products such as oil and gas. Oil giants top the list, led by Saudi Aramco with a 4.33 per cent share, followed by Russia's Gazprom (3.17), Chevron (3.10), ExxonMobil (3.01), National Iranian Oil Co. (2.62), BP (2.45), Royal Dutch Shell (2.30), Coal India (1.73) and Mexico's Pemex (1.63). PetroChina is in tenth place with a 1.17 per cent share. RWE is well below the top twenty.

"The RWE case is unique with respect to requesting a symbolic judgement based on the company's share of global emissions," Heede told me in an email. "But the case is not unique in terms of complaints and litigation against fossil fuel companies: there are now some two dozen cases, chiefly in the US."

Heede originally measured corporate emissions since the Industrial Revolution began in the eighteenth century – long before anyone was aware of global warming. He has since shifted the baseline to 1965, arguing that "recent research has revealed that by the mid-1960s the climate impact of fossil fuels was known by industry leaders and politicians".

That year, US President Lyndon Johnson told Congress,[35] "This generation has altered the composition of the atmosphere on a global scale through ... a steady increase in carbon dioxide from the burning of fossil fuels." Published also that year, a White House report[36] on 'Restoring the Quality of our Environment' was ahead of its time in suggesting that the "melting of the Antarctic ice cap" might be one effect of an increase in carbon dioxide in the atmosphere. "The melting of the Antarctic ice cap would raise sea level by 400 feet," it said. "If 1,000 years were required to melt the ice cap, the sea level would rise about 4 feet every 10 years, 40 feet per century. This is 100 times greater than present worldwide rates of sea level change."

However, like RWE, many nations reject the idea that they could be liable for individual impacts of climate change. The 2015 Paris Agreement,[37] which seeks to phase out global emissions in the second half of this century to limit warming, does not include the word "liability". The US, under former President Barack Obama, was among big emitters who insisted on leaving out any suggestion of blame for the climate crisis. As part of a delicate balance between governments, it also has no sanctions on nations that fall short of their promises to cut emissions. Instead, it is based on a fuzzy, friendly approach that allows each of almost 200 nations to set its own goals. Compliance with promises to curb emissions is to be overseen by a committee of experts "and function in a manner that is transparent, non-adversarial and non-punitive". Such wording of 'non-punitive' is hardly a green light to encourage lawsuits.

Despite evidence that governments and companies were aware of the risks of global warming decades ago, the IPCC only began tentatively blaming human emissions in 1995. That link has since grown much more robust. But there is still a gulf between the scientific and political consensus and proving to a court that RWE's emissions are a cause of the thaw of Peru's glaciers. Plus, the issue of historical liability is a can of worms. Is Brazil, for instance, liable for the decades of deforestation of the Amazon that released vast amounts of greenhouse gases to the atmosphere? Is Beijing to be held to blame for farmers who unwittingly emitted methane, a powerful greenhouse gas, by planting rice in paddies? Is Britain liable for its emissions, and those of its former colonies, back to the start of the Industrial Revolution in the eighteenth century?

Everyone I met in almost a week in the Huaraz region said that the snow cover was receding – unsurprisingly. Visit the Pastoruri glacier a few tens of kilometres east of Huaraz at 5,200 metres and there are signs placed hundreds of metres apart down the bare mountainside showing where the glacier stood in previous decades, including pictures of skiers on what is now barren ground. At the top, the current end of the shrunken glacier looks like a meringue spattered by rain, sleet and mud. A large chunk has detached from the end of the glacier, leaving a crack like a giant frown at the snout.

The current melt of mountain glaciers is now a pulse of extra water towards the sea, but that will dry up fast in coming decades. Worldwide, melting glaciers outside Antarctica and Greenland are projected to raise sea levels by between about 7cm and 20cm by 2100, according to the IPCC in 2021.[38]

In China and India hundreds of millions of people rely on the flows of rivers from the Ganges to the Yangtze that rise in the Himalayas. In the dry seasons, melting glaciers provide a steady stream of water for irrigation. In regions with mainly smaller glaciers, such as central Europe, Scandinavia, the United States, north Asia, and New Zealand, glaciers will lose more than 80 percent of their ice by 2100 with high emissions, the IPCC said. And it said the global retreat of glaciers was "unprecedented" in at least the last 2,000 years. That will cause massive disruptions, from irrigation in Tanzania to Alpine ski resorts.

The outlook is bleak for places like Palcacocha.

"Glacier retreat and permafrost thaw are projected to decrease the stability of mountain slopes and increase the number and area of glacier lakes," the IPCC says. "Resulting landslides and floods, and cascading events, will also emerge where there is no record of previous events."

In 2003 in Huaraz, an alarm went off from Lake Palcacocha after a landslide into the lake sent waves of water over the dam, triggering a call to evacuate. Luckily the flood was not big enough to reach the city.

"People rescued their families – and carried out their TVs," said Beltrán, who guided me up to Palcacocha, laughing at the thought of what the locals valued most after their children, spouses and other relatives.

People in Huaraz are regularly warned their city is under threat from melting ice far away up the valley – there is even a big poster about Palcacocha in the bus station showing parts of the city most vulnerable to flooding in red as well as evacuation routes uphill, away from the river. Visitors can check on arrival where they're staying on the map and make a mental note of where to run if the siren goes off. Indeed, Palcacocha is so much a focus of worries in the high Andes that news about it is sometimes misleading, spooking residents and scaring off would-be tourists vital for local hotels and restaurants.

In April 2003, NASA released a statement[39] warning of an ominous crack in the glacier above the lake, based on satellite photos. "A fissure has appeared in the glacier that feeds Lake Palcacocha," it said. "If the piece breaks off, ensuing floods would take 15 minutes to reach the city. In 1941, the lake overflowed and caused massive destruction, killing 7,000 people."

"The press release sent shock waves through Peru as worried residents, scientists, and authorities demanded details about their fate – and tourists cancelled their plans to visit Huaraz," according to a report by Peru's government-run National Institute for Investigations of Glaciers and Mountain Ecosystems (INAIGEM), based in Huaraz. Unnerved by the reports of a crack, only 6,000 people came to Huaraz for Easter celebrations that year, for instance, against a projected 18,000, according to the Peruvian daily *El Comercio* at the time.

Fortunately, NASA got it wrong – Peruvian experts were quickly sent to inspect the dark line on the glacier from the ground to check the satellite measurements. "The supposed crack in the glacier was a rock," INAIGEM said.

In Huaraz, I visit retired engineer Cesar Portocarrero, a renowned expert who has championed projects to tame Peru's Andean lakes. He argues that Palcacocha has been a perennial problem and that it might be better to drain it faster, as has been done with some other lakes, by building tunnels to allow water to flow out under the dam, rather than rely on siphons. "Palcacocha has glaciers hanging above it. If these fall into the lake it will overflow," he says. Portocarrero says the best way to save glaciers is to reduce greenhouse gas emissions worldwide. But he's sanguine about how much pressure nations such as Peru, which have done little historically to emit greenhouse gases compared to rich industrialised nations, can exert on others to slow global warming. "The only ones who will cope with climate change are those with a lot of technology, a lot of money. The Netherlands has problems with rising seas, but they have money. The poor countries have to adapt. How are we going to struggle against China, the United States, the European Union to get them to act? We can't fight – we have to adapt. ... And the most important element is water. Water is being most affected by climate change ... Civilizations have disappeared because of a lack of water. If we don't take measures these beautiful glaciers in the Cordillera Blanca will disappear."

He's a gentle man, and proudly shows a large silver medal, hanging from bright blue, white, red, green and yellow ribbons, that he was awarded in 2016 by a Nepalese non-governmental organisation for his expertise on averting glacial lake outburst floods. The disc – a Sir Edmund Hillary Mountain Legacy Medal – is awarded once every year or two for "remarkable service in the conservation of culture and nature in mountainous regions". It even has its own wooden box, locked with a tiny key. He laughs that the key seems unnecessary since anyone wanting to steal the medal could simply swipe the small box.

Among weirder attempts to limit glacier melt, Portocarrero noted that in 2009 the World Bank backed an experiment further south in the Andes to paint an entire mountainside white, hoping that it would reflect more sunlight back into space, cool the region and allow the regrowth of glaciers. The mastermind of that massive whitewash project was Peruvian entrepreneur Eduardo Gold. After some initial work, however, it proved impossible to slop the whitewash over enough of the Chalón Sombrero Mountain near Ayacucho – and rains just washed off the white, made from lime, industrial egg white and water. Gold, who has since passed away, told the BBC at the time, "Cold generates more cold, just as heat generates more heat." "I am hopeful that we could re-grow a glacier here because we would be recreating all the climatic conditions necessary for a glacier to form," he added. Portocarrero said the idea was quixotic at best. "It's very nice as an experiment but it's impossible in the Cordillera. Gold painted some mountainsides – but imagine trying to paint the entire mountain range – it's crazy." And the whitewash itself disrupts local plants and animals.

Perhaps even stranger, there have been experiments to spread sawdust on snow in the Andes as an insulating blanket – something attempted in the Alps where snow is valuable, for example on the lower slopes of ski resorts where tourists want to ski to their hotels. But Alpine resorts are often set in pine forests. By contrast, trucking sawdust high into the Andes where there are no trees is all but impossible, Portocarrero says. The emissions from all the trucks and work in cutting down trees and creating sawdust would be massive. There has been one project to lay down sawdust – a layer of about 15cm – by Benjamin Morales Arnao on the Pastoruri and Chaupijanca glaciers in Peru.

Alongside risks for the ice itself, global warming may be making the Andes more dangerous for climbers and guides such as Luciano Lliuya and Beltrán. They both said the thawing snow is making it harder to judge treacherous ice. Neither can say off the top of his head how many times he has climbed to the top of the iconic Alpamayo peak – a steep pyramid north of Huaraz dubbed "The Most Beautiful Mountain in the World" by German magazine *Alpinisimus* in 1966 – but they reckon it's at least twenty times each.

Alpamayo looks like the jagged Matterhorn in the Alps, but is higher with a peak at 5,947 metres. And these are people who really gauge and understand risk of changing weather and climate in their everyday lives. Beltrán says it's best, for instance, to be first leaving a tent camp for a final assault on a peak like Alpamayo in the middle of the night when it is coldest. Also, it helps to be first if there are other teams climbing because there is a risk that less-experienced climbers will dislodge chunks of ice onto those below. Every year there are fatal accidents here in the Andes, and the guides fear the risks are growing as temperatures rise. "This year we met the New Year on Nevado Mateo 5,150 metres," Beltrán wrote on 1 January 2021. "Not such a favourable climate, but all well."

So, working as an Andean guide demands a determination and deep understanding of the risks of thawing ice in a changing climate. In Huaraz, Luciano Lliuya says he'll apply a similar resolve to the RWE case, even though it could take years, and may ultimately fail. "When you see these changes in nature something has to be done. You can't just wait for things to get worse, for the mountain snows to disappear ... We just have to try."

4

IN MONACO, SUPERYACHTS AND SEA LEVEL RISE

> We need to take immediate and drastic actions to cut emissions right now ... if we want to achieve carbon neutrality by mid-century.
>
> IPCC Chair, Hoesung Lee

In 2021, scientists released their most grim report yet about global warming, saying it is now "unequivocal that human influence has warmed the atmosphere, ocean and land" and that changes to the ocean, ice sheets and global sea level may be irreversible for centuries or millennia. The foundations for these findings about the ocean and ice were laid in 2019 when scientists met in Monaco, against a backdrop of high-emitting sports cars and luxury yachts.

SUPERYACHTS DOT THE AZURE MEDITERRANEAN, a sleek black sports car gleams in a showroom labelled as a 'bargain' at €288,000, and a bijou studio apartment with a view of the bay here in Monaco is on sale for €2.2 million.

A few hundred metres down the hill from the Monte Carlo casino, climate scientists and officials from more than 100 governments are a world away, working on a UN report about the dire, and worsening, outlook for oceans and melting ice. They've come to complete a bombshell study that includes warnings about how emissions from burning fossil fuels are exacerbating risks of abrupt changes for the planet that could herald a collapse in fish stocks or trigger a runaway melt of Antarctica's ice sheet.

The conclusions are obvious: the world needs to shift from high-carbon lifestyles – exemplified by the super-rich here in low-tax Monaco – to limit harm to the ocean and a sharp rise in sea levels.

It's September 2019, and the delegates have gathered in the principality with almost 40,000 inhabitants for a five-day meeting in a building among palm trees near the waterfront, close to where Formula One cars whizz past during the annual Grand Prix. The building, with sloping sides and a flat roof, looks like a half-completed giant glass pyramid.

This is the final lap to complete two years of work by more than 100 scientists from thirty-six nations in the IPCC, the UN's authority on global warming. They've distilled more than 7,000 scientific reports into a draft summary for policymakers of about thirty pages. The job now is to go word by word through the text and approve what the IPCC calls its 'Special Report on the Ocean and Cryosphere in a Changing Climate' (SROCC).[1] It all sounds dauntingly academic – cryosphere is from the Greek word *kryos* meaning 'cold' – but the dryly phrased draft conclusions show the world is far from cold and is at a planet-changing fork in the road in safeguarding the oceans and limiting sea level rise.

In the run-up to the meeting, the French news agency AFP has obtained a leak of the draft for a story headlined 'Oceans turning from friend to foe, warns landmark UN climate report'. It starts, 'The same oceans that nourished human evolution are poised to unleash misery on a global scale unless the carbon pollution destabilising Earth's marine environment is brought to heel.'[2]

The draft shows that climate change will add to pressures on fish stocks and acidify the oceans as carbon dioxide mixes with the waters. But it also says all is not yet lost, a message that IPCC reports have made relentlessly for years – and repeated in 2021 even as the outlook has grown darker.

In 2021, the IPCC[3] said its findings about future sea levels were "broadly consistent" with the Monaco report. Both say that seas could rise about a metre this century in the worst likely case of rising emissions. The 2021 report projects that seas could rise by between about 2 and 7 metres by 2300, above the 2.3 to 5.4 metre range in Monaco, although the two are not exactly comparable because of different scenarios. Slash emissions and both reports show seas will rise more slowly, less than a metre by 2300 in the best case. That would be awful, but probably just about manageable with three centuries of human ingenuity.

In one big change, the 2021 report introduces a note that it "cannot rule out" far higher rises of sea levels by 2300 if ice sheets on Antarctica and Greenland disintegrate. In 2019, those fears were known, but too uncertain to merit mention.

In Monaco, scientists know that getting the report to what some call the chequered flag of approval – scientists make weak Formula One jokes at the meeting – will be hard because of the political pitfalls at IPCC meetings. Until now, scientists around the world have been doing their usual day jobs: collecting data from ice sheets, from high mountain glaciers, examining fish catches and measuring the state of coral reefs. Here in Monaco things are different as they abruptly run into politics.

That's because the IPCC is an awkward hybrid organisation in which the scientists' work has to be approved by governments including the poorest African nations, the Organization of the Petroleum Exporting Countries (OPEC) and top emitters led by

China, the US and the European Union. That twin approach is meant to give the reports more weight as the basis for future policies, but government delegates at the talks can turn up with their own agendas, often unrelated to science. Among the scientists' top fears is that the US is run by President Donald Trump, and he is pulling the US out of the 2015 Paris Climate Agreement – a decision that was to be reversed by President Joe Biden upon taking office in 2021. Trump has frequently tweeted his scepticism that global warming is man-made, even calling it a hoax invented by China to harm the US.

Elsewhere, Saudi Arabia and its OPEC allies, sceptical about acting boldly to limit global warming, don't want to undermine their oil-dependent economies. They often send officials affiliated with the nation's top oil company, Aramco, to IPCC meetings. At the other end of the spectrum, in IPCC terms, are the poorly funded delegates from small island developing states desperate for more action to limit greenhouse gas emissions. Fourteen island states first warned in the 1989 Malé Declaration,[4] named after the capital of the Maldives in the Indian Ocean where the meeting was held, that climate change and rising seas could "threaten the very survival" of some nations.

Given all this, some of the deepest political and ideological rifts in the world come to the fore at IPCC meetings. There are a lot of nervous delegates at the Monaco meeting when it kicks off, though you wouldn't know from the polite, restrained tone. "This week's deliberations on the Special Report on the Ocean and Cryosphere in a Changing Climate marks another major milestone for the IPCC," IPCC Chair Hoesung Lee of South Korea intones at the opening on 20 September.

Scientists work for the IPCC for free – it's about prestige, not pay. The IPCC shared the 2007 Nobel Peace Prize[5] with former US Vice President Al Gore. Since then, many scientists have added a Nobel logo on their personal webpages, even though it's a very diluted share of glory when the IPCC comprises thousands of contributors over the years. While IPCC work is unpaid, scientists get their travel costs and expenses covered to often exotic venues like Monaco, before the pandemic pushed everything online in 2020. But the scientists know they have to get it right. In 2010, the IPCC had to issue an embarrassing correction to a 2007 report that mistakenly said that Himalayan glaciers could all melt away by 2035.[6] That projection of a shockingly fast loss, with implications for hydropower and irrigation in Asia, was simply caused by poor research. The IPCC has tightened checks, but errors are a nagging worry for the authors.

Adding to the fractious politics at the IPCC meeting, there are massive uncertainties in the report about climate change. The future of ice sheets in Antarctica and Greenland, which lock up ice equivalent to 58 metres and 7 metres of sea level rise respectively, is one of the areas of climate science most fraught with ambiguity. And somewhere there is a temperature tripwire, a tipping point beyond which the ice locked up in the two ice sheets starts an unstoppable thaw.

Can that tipping point be identified? How hot does the planet have to get? Is an irreversible melt already under way? Some scientists fear it is, others are doubtful. And what does 'irreversible' even mean? Will ice shelves and 'ice cliffs' around Antarctica collapse into the sea and accelerate the melt? As mentioned before, some climate scientists liken glaciers at the end of ice sheets to a bottle of wine on its side – as long as the cork stays in, everything's fine. Take away the cork and most of the bottle will spill. By that analogy, Antarctica's glaciers are giant bottles of wine with the corks pointing out to the coast and the Southern Ocean. At the other end of the Earth, Greenland's ice is melting ever quicker.

Two years after the Monaco meeting, the 2021 report shows the IPCC is still wrestling with the pace of the thaw. But it is now certain that we are responsible for climate change and it says "continued ice loss over the 21st century is virtually certain for the Greenland ice sheet and likely for the Antarctic ice sheet".

In Monaco, participants are banned from discussing the debates in the hall with outsiders like me, so it has to be done furtively in cafés or bars nearby, past the supercar showrooms. One of the participants agrees to meet near the main building one evening, up an outdoor staircase where we can't be spotted from the entrance.

Getting up high in the evening sun has the advantage of opening up a spectacular view of the bay and the glittering yachts, invisible from ground level. At the entrance to the conference hall, a building site blocks the view of the sea for delegates trying to imagine the fate of the unseen, future ocean.

"The Saudis are by far the most vocal here and they're very good at this ... It's all about undercutting the report's conclusions," she says, asking not to be named. She and other delegates say the Americans, by contrast, are sticking to their recent run of not standing in the way of research findings – many of the US scientists are dismayed by Trump's scepticism about climate science and simply refuse to play along. The realisation that the Americans are not trying to block the meeting is a big relief to most other delegations. "Every time you turn up to these meetings you fear the Americans will be enforcing Trump's line. Here they're not," one delegate said.

It's a complex, wily game. Like UN meetings on climate change, they always go down to the wire, with often Byzantine debates. The placement of a single comma in a UN document famously took up two hours in Bali, Indonesia, in 2007. The IPCC delegates know they are in for a sleep-sapping few days. IPCC meetings almost always run overnight at the end. Some delegates store up key demands for changes to the text to spring them at the last minute, hoping they will be accepted simply because many other delegations are too exhausted to care.

The late IPCC Chair Rajendra Pachauri, a keen cricketer from India, used to warn delegates not to count on his weariness to get their views across in the depths of overtime in the final night. Pachauri and I went to an Indian restaurant one evening in 2012 at his part-time residence in New Haven, Connecticut, where he was affiliated to Yale University. He said it was important for the chair to show that she or he was in charge. He said he once told an IPCC meeting at the start, "I will outlast you. I can happily stay up two nights in a row without sleep and then go out and bowl out half of the other team before lunch." He shrugged, "You just have to accept" the disruptive tactics of delegations who see themselves as defending their national interests.

The Saudis are historically vocal, both for their own interests and also as proxies for climate doubters such as Brazil's right-wing populist President Jair Bolsonaro. Saudi delegates at IPCC and other UN climate meetings consistently say they are merely trying to clarify, not obstruct, the science. In Monaco, Saudi Arabia's delegation, led by Ayman Shasly, raises repeated objections. Sometimes they win, sometimes they lose.

A detailed account[7] of the session subsequently published by Earth Negotiations Bulletin (ENB), run by the International Institute for Sustainable Development which was allowed to track the closed-door talks, mentions Saudi Arabia forty-seven times as raising points about the text. No other nations come close in the ENB account: France lags on thirty-two mentions, while nations including the US, India and Britain, and Saint Kitts and Nevis – often speaking on behalf of small island states – are all in the twenties.

According to ENB, at one point in discussing marine heatwaves – bursts of abnormally warm water that can kill coral reefs – Saudi Arabia says it would be better to drop any mention that they are caused by humans. Other nations object, and the text ends up saying they are very likely caused by human activities. Saudi Arabia plays a long game at IPCC meetings, often pedantic but rarely blocking decisions that have to be adopted by consensus. The Saudi delegation takes an odd dual approach, insisting that it takes climate change seriously while blaming global demand for oil for its economic dependence on fossil fuels. Crown Prince Mohammed bin Salman said in 2021, for instance, that the nation will generate 50 per cent of its domestic energy from renewables by 2030 and its people suffer from desertification, dust storms and air pollution from fossil fuels. Saudi Arabia's promotion of solar power at home has the coincidental effect that it means more oil to export.[8]

Shasly was asked in a 2018 interview by Carbon Brief[9] what he would tell vulnerable, low-lying states who want a radical shift from the fossil fuels OPEC champions. "I'll tell them: 'Go and speak to the consumers, the large consumers, those who are emitting the most, those who are impacting the climate. You must talk to them. Tell them: what have you done over the last 200 years?'"

In the end, Riyadh and its allies come away with at least one semantic victory, even if it looks pretty meaningless at first glance. The final text says the SROCC report "follows" other UN reports over the previous year about the risks of temperature rises of 1.5°C and climate change and land, as well as a separate UN report about threats to biodiversity that showed a million species are at risk of extinction.

Many other nations want the SROCC to stress that the reports are a family that all address facets of humanity's impact on life on Earth. They favour saying that the SROCC 'complements', 'expands on' or 'reinforces' the other reports. That would put more pressure on governments to deal with twin crises of biodiversity and global warming as part of the same overarching human threat to the planet. Riyadh gets its way in the end partly because other delegates say there is reluctance to fight too hard. Lurking in the background is the fear that it could open up far bigger cans of worms about core issues that might draw in Washington and Trump's scepticism.

Still, oil exporters can now say that the wording that the SROCC 'follows' other reports makes it a mere fact of chronology, toning down any link by uniting unconnected events in the same way as saying that the Monaco Grand Prix in May 2019 followed a Nigerian general election in February the same year.

Everyone knows that Riyadh is a formidable negotiator.

The heart of the 2015 Paris Climate Agreement, for instance, is widely seen as ending the world's addiction to fossil fuels and shifting to cleaner energies such as solar and wind power. But you would hardly know it from the text. The Paris Agreement refers only to 'greenhouse gases' with no mention of 'oil', 'coal', 'natural gas' or 'fossil fuels'. Oil producers of course like to stress that there are other, smaller sources of greenhouse gases – such as methane from cows' digestive tracts, or from rotting vegetation or bubbles from rice paddies – even though carbon dioxide from burning fossil fuels is the main heat-trapping gas.

The IPCC is sometimes criticised as being too conservative because of the demand for consensus that requires the backing of nations including reluctant OPEC oil producers. The first time the IPCC tentatively blamed humankind for causing climate change was in a report in 1995 that concluded in twelve famous words: 'The balance of evidence suggests a discernible human influence on global climate.' That sentence was a revolution by starting to point the finger of blame at humanity, a link that the IPCC upgraded to "unequivocal" in 2021. Accepting that human influences are the dominant cause opens the door to the economy-changing question: 'What are you going to do about it?'

In his book *Science as a Contact Sport*,[10] late US scientist Stephen Schneider said that Mohammed Al Sabban, Saudi Arabia's climate negotiator in the mid-1990s, and OPEC nations wanted to scrap the entire chapter that concluded there was a 'discernible' human impact. Schneider wrote he favoured letting the Saudi objections be known publicly. He felt that if this happened, 'They will become a world laughing stock, these OPEC special interests, claiming to know more science than the lead authors and the rest of the world.' Saudi Arabia backed down and let the language pass at a meeting a month later.

Yet the famous sentence that 'the balance of evidence suggests a discernible human influence on global climate' provoked a torrent of abuse from defenders of fossil fuels for Ben Santer, a scientist at the Lawrence Livermore National Laboratory in the US, who came up with the phrasing. Santer told an interviewer in 2019 that a man left a dead rat outside the door of his house and yelled abuse as he drove away. He tried to shield his family from knowing too much but his young son picked up on the worries and started to sleep with a wooden sword under his bed – just in case.[11]

Most oil producers have long since bowed to the overwhelming scientific evidence that fossil fuels cause climate change, and Saudi Arabia ratified the Paris Climate Agreement in November 2016. Still, Saudi Arabia has proved itself brilliant at arguing that maybe the risks about future temperature rises are exaggerated, and that exporters need help to adapt.

In Monaco, most of the debate is less explosive, mostly about how best to phrase the conclusions to make them clearer. Small island states persuade other delegates to add them to a list of human communities "particularly exposed" to changes in the ocean and the cryosphere, along with a broad mention of those living by the coast, in polar areas and high mountains. Indigenous peoples and Indigenous knowledge gain a capital 'I' in the report, shifting from the previous style of a lower-case 'i'. The change is to give Indigenous peoples more respect.

In a final spurt to get the report finished, delegates stay up all night on the final day, often breaking off into smaller groups to sort out disputes. IPCC reports, and UN sessions about climate change, are unusual because they have to end with 'consensus'. That is a deliberately vague yardstick which stops short of 100 per cent agreement and does not give individual delegations a veto. There is no vote to approve the report at the end, no show of hands. Reports are adopted when the IPCC chair judges that there is 'consensus', banging down a gavel once there are no lingering hands raised with objections around the hall. In this case, that happens at about 10.30 in the morning on Tuesday after a marathon overnight session, prompting a standing ovation for the thirty-six-page SROCC summary for policymakers. The four co-chairs of the meeting, Debra Roberts of South Africa, Hans-Otto Pörtner of Germany, Valérie Masson-Delmotte of France and Panmao Zhai of China, stand and raise their arms in triumph.

The report confirms the risks of sea level rise to 2300 in the draft, underscoring that action now is essential to limit what could be several metres of sea level rise if emissions keep climbing. And weary delegates say that, overall, the core messages were undiminished by the behind-the-scenes wrangling.

"The report came out stronger," said Martin Sommerkorn, a scientist and expert on the Arctic at the meeting who works for the WWF conservation group. "We see a new environment coming," he told me. Sommerkorn lamented he had to go forty hours without sleep until the 10.30 conclusion. A former top German decathlete, he is lucky to have the stamina to match his rivals.

At a news conference the next day, scientists present the findings at Monaco's Oceanographic Museum where live turtles swim in an aquarium and a stuffed polar bear gazes down from a glass case. The start has been slightly delayed because Prince Albert wants to attend but has to arrive on an overnight flight from the US.

Lee, the mild-mannered IPCC chair, is part of a panel of eight IPCC scientists sitting behind a long wooden desk in the museum, on a promontory high above the Mediterranean. He is unusually blunt about the implications of the report and the risks of rising seas. "We need to take immediate and drastic actions to cut emissions right now ... if we want to achieve carbon neutrality by mid-century," he tells the audience, including Prince Albert.

Afterwards, some climate scientists sitting at the back of the room said they murmured in surprise at Lee's call for "immediate and drastic actions" – Lee is usually more circumspect, they said, and the IPCC's mandate is to do the science, not prescribe policies.

Pachauri, the former IPCC chair, was often accused of crossing the line into politics. He once said, for instance, that he hoped that a previous IPCC report would "shock the world" into action. That made him unpopular with some major emitters, who felt they were being accused of inaction, while endearing him to vulnerable nations. Pachauri, who died in 2020, raised the IPCC's profile around the world. He quit as chair in 2015 after a female employee at his research company accused him of sexual harassment, a charge he denied.

An introductory video about the Monaco report shows threats to the oceans and ends with the message: "Our future is in our hands." Even setting aside the crisis of rising sea levels, the ocean is suffering like never before from over-fishing, plastic pollution and acidification caused by greenhouse gases, it says. Greenhouse gases form a mild acid when they mix with water, and that is making it harder for animals such as oysters, crabs, lobsters and corals to build their shells. And less than 10 per cent of the seas are in protected areas where safeguarding life is the priority.

The IPCC statement summing up the findings is marginally less stark than Lee's call for action. It starts:[12]

> Choices made now are critical for the future of our ocean and cryosphere.
> MONACO, Sept 25 – The latest Intergovernmental Panel on Climate Change (IPCC) Special Report highlights the urgency of prioritizing timely, ambitious and coordinated action to address unprecedented and enduring changes in the ocean and cryosphere.
>
> The report reveals the benefits of ambitious and effective adaptation for sustainable development and, conversely, the escalating costs and risks of delayed action. The ocean and the cryosphere – the frozen parts of the planet – play a critical role for life on Earth.

Among the report's main findings about sea level rise are that 680 million people, almost a tenth of humanity, live in coastal areas less than 10 metres above sea levels. And the number is on track to exceed 1 billion by 2050.

Among the bleak findings is that floods along the coast will become ever more frequent as seas rise, no matter whether governments cut greenhouse gas emissions or not. That means increasing coastal inundations, ranging from ports to farmland.

"Extreme sea level events that are historically rare (once per century in the recent past) are projected to occur frequently (at least once per year) at many locations by 2050" in all scenarios of greenhouse gas emissions, it says. It includes a map of the world showing dots around the coasts of the most vulnerable sites, many of them in the tropics.

And it gives an overview[13] of where the ice is already melting – seas rose by 3.6mm a year in the decade to 2015. The biggest single contributor in the period, 1.4mm a year, is simply because the water in the oceans swells as it gets warmer, a process known as thermal expansion. Other big contributions are Greenland (0.77mm), Antarctica (0.44mm) and other glaciers around the world (0.61mm), with other variations from water stored on land. Each set of measurement comes with uncertainties. "Sea level at the end of the century will be higher than present day and continuing to rise in all cases even if the Paris Agreement is followed," the report says. The 2021 report revises up the accelerating rise in sea levels by a fraction, to 3.7mm in the period from 2006–18, while broadly reaffirming the various sources of the melt. In the worst likely case, seas could be rising by more than 15mm a year by the year 2100.

The Monaco report points to uncertainties, especially about the current thaw of ice sheets in Antarctica and Greenland, and the risks of an unstoppable melt: "Acceleration of ice flow and retreat in Antarctica, which has the potential to lead to sea level rise of several metres within a few centuries, is observed in the Amundsen Sea Embayment of West Antarctica and in Wilkes Land, East Antarctica." And it adds: "These changes may be the onset of an irreversible ice sheet instability." That sentence includes a massive threat – "irreversible instability" is losing the cork from the proverbial wine bottle lying on its side that could lead to the disintegration of ice sheets. A footnote also adds that "irreversible" means losses over centuries to thousands of years.

In 2019, several countries, small island nations and some European countries, such as the Netherlands, stressed the importance of a clear message about the risk of a planet-changing melt, according to the account by ENB. Others were reluctant to go too far in highlighting even the possibility of a melt that would change maps of every continent.

A mere two years later, based on more evidence, the IPCC sounded the alarm about the risks of an irreversible disintegration of Antarctica, adding the note that sea level rise of more than 15 metres "cannot be ruled out" by 2300. And the summary of the 2021 report said: "many changes due to past and future greenhouse gas emissions are irreversible for centuries to millennia, especially changes in the ocean, ice sheets and global sea level." (See the plate section of this book for projected rises of sea levels in the 2021 report.)

In predicting the future, the 2021 report stresses that ice sheets can fall apart quickly but, once they vanish, take far longer to regrow. It also warns that "reducing atmospheric carbon dioxide concentrations or temperatures to pre-industrial levels may not be sufficient to prevent or reverse substantial Antarctic mass losses."

Both the 2021 and 2019 reports conclude that humanity has set the Earth on track for massive changes.

In Monaco, Masson-Delmotte, one of the co-chairs of the report, tells the news conference, "The planet will experience sea level rise for decades and centuries to come".

Luckily, no reporters ask what the IPCC leaders think about the message the report sends to the Monaco Yacht Show, extravagant even by the local standards of luxury. The show's biggest yacht is the 111.5-metre-long *Tis*, with space for eighteen guests and thirty-eight crew, and a lavish list of luxury features including heated marble floors, two helipads, a gym, a lift, swimming pool and a wine cellar. Lürssen, the German shipyard which built *Tis*, says it is "the largest yacht ever seen at a yacht show". One website quotes a price of €2.2 million to rent the vessel for a week.[14]

Prince Albert, sitting in the front row, does not speak at the IPCC event.

I once interviewed Prince Albert at a meeting about threats to nature in Barcelona in 2008 – he candidly said people sometimes asked him when he urged more action on climate change: "How can you speak about these issues when you have a Formula One race in your backyard?" In the long run, he said, he favoured running the Grand Prix on biofuels. "That's the aim, it will take some time but I'm sure we will get to that," he said. It hasn't happened yet. But he does generally get good marks for trying to curb emissions – he was once named a "Champion of the Earth" by the UN Environment Programme.

Still, Monaco has managed in the past to have some imaginative accounting to cover over emissions from the super-wealthy. In 2003, Monaco estimated that its greenhouse gas emissions were up by a shocking 37.8 per cent from a 1990 benchmark, despite a commitment to curb them. The next year, Monaco revised them down to a laudable fall of 3.1 per cent since 1990. Monaco had taken advantage of a rule that fuel used by aircraft and ships on international trips are excluded from national greenhouse gas emissions. Monaco decided that almost all yachts leaving the harbour will cross into French or Italian waters just outside the bay, so those emissions don't count. Also, Monaco estimates that most helicopter flights are to Nice airport or elsewhere abroad – less than a fifth of flights are 'domestic'. Suddenly, thousands of gallons of fuel don't count towards local emissions. Now, Monaco can boast of a 15 per cent decline in emissions[15] by 2018 since 1990, also helped by greener policies for energy efficiency and renewable energies pushed by Albert.

As always with climate change, almost no one is doing enough.

In Oslo, a few weeks after the Monaco meeting, I again meet Pörtner, one of co-chairs of the SROCC report. Making a speech about the SROCC, he displays a PowerPoint photograph of two men and a woman kneeling on a beach with their heads buried in the sand. They look like officials, with suits or a dark formal dress. And he has added the caption, 'A common response even among those who (should) know ... !?'

That doesn't always go down well, he tells me.

"It's a challenge for some to see. I once showed it in the European Parliament and I got a rather emotional response [...] They felt accused of not doing enough", he smiled.

The 2021 IPCC report broke new ground by being the first negotiated and approved online, a tribute to scientists working unpaid for the IPCC at home amid the pandemic, often with sporadic wifi connections, especially in developing nations. And with people calling in from around the world, it was always the middle of the night for someone.

But it was not agreed without often fierce, if polite, debate. A record of the meetings by ENB[16] showed that Saudi Arabia, for instance, was among nations objecting to the conclusion that humanity's greenhouse gas emissions have "unequivocally" caused warming. Saudi Arabia favoured vaguer language that would be less of an indictment of humanity.

Saudi delegates, however, eventually agreed to a final phrasing that it is "unequivocal" that "human influence" has warmed the atmosphere, ocean and land. Notably, that stops short of mentioning fossil fuels – coal, oil or natural gas – the main source of carbon dioxide emissions.

Like Monaco where some scientists fretted that the IPCC chair Lee was straying into policy by saying that "drastic actions" were needed to cut emissions, Russia, China, South Africa and Saudi Arabia also objected to some phrasing in the report urging radical reductions. They finally to a final headline statement that the goals of the Paris Agreement require "deep reductions" in greenhouse gas emissions.

So, two years on from Monaco, scientists say calls for drastic action are ever more mainstream. "This is physics. The more emissions, the more warming. This has nothing to do with policy prescription," Gudfinna Adalgeirsdóttir, professor of glaciology at the University of Iceland and an IPCC lead author of a chapter about melting ice, told me of the 2021 report.

5

FUTURE MIGRATION: A FLAWED PACIFIC GUIDE

They were not listening, simply sitting and repeating to themselves, 'We want to keep our land'.

British colonial official Arthur Grimble in 1928, trying to persuade Banabans to sell their Pacific island to allow phosphate mining

The Pacific has a unique moral authority to speak out. It is time for the world to listen.

UN Secretary-General António Guterres

In 1945, about 1,000 people from Banaba island in what is now the Pacific island nation of Kiribati were forced to move to Fiji after British colonial authorities devastated their homeland with mining. As global sea levels rise, low-lying island nations fear they may have to migrate – the Banabans are a rare historical example of a people torn between two homelands, with seats in the parliaments of two nations.

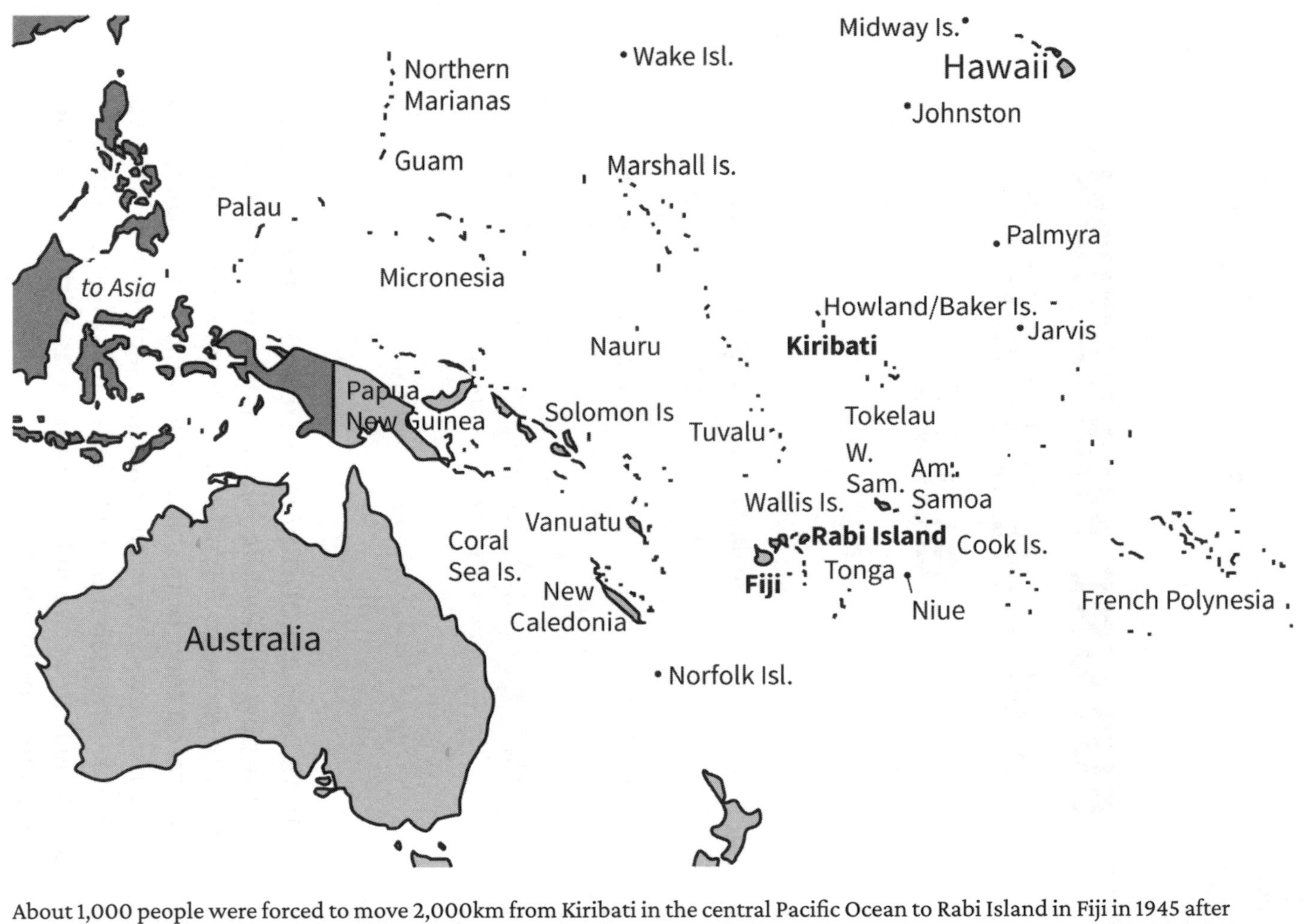

About 1,000 people were forced to move 2,000km from Kiribati in the central Pacific Ocean to Rabi Island in Fiji in 1945 after British phosphate mining ruined their homeland.

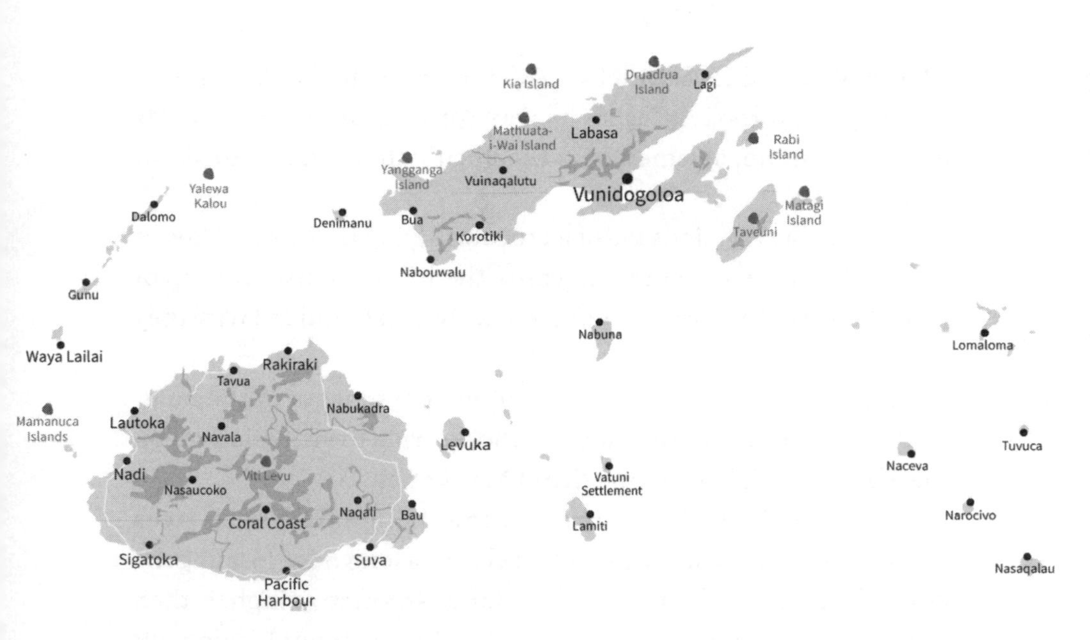

The Pacific island nation of Fiji, showing Rabi Island and Vunidogoloa village (both upper right). In 2014, Vunidogoloa was the first village moved inland by Fiji to escape coastal flooding (see Chapter 2).

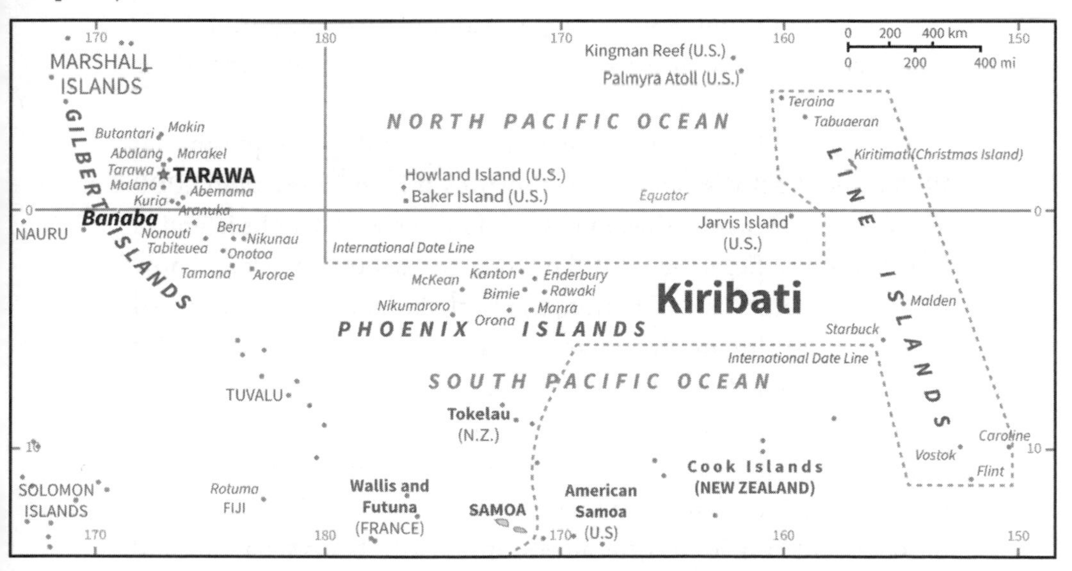

Kiribati, a nation of thirty-three islands in the central Pacific with the capital, Tarawa. Banaba Island is the ancestral home of people relocated to Rabi Island in Fiji in 1945. Ioane Teitiota, who lost a bid to become the world's first "climate refugee" in New Zealand courts (see Chapter 10) lives on Tarawa.

THE PASSENGERS BLINK BACK EYE-STINGING spray as the small motorboat splashes its way through the waves towards Fiji's Rabi island, an emerald in the Pacific that is home to about 5,000 people.

"Rabi always reminds me of a crocodile lying in the sun," Rosite Iotua tells me as she points towards the island, a distant strip of green between the dark blues of the sea and a sky studded with grey rain clouds.

In her mid-twenties, she's one of six passengers in the boat, smiling, chatting and enjoying the choppy ride to visit relatives on the island, which is about the size of Manhattan.

Closest is the tail, says Iotua, a low-lying forest with trees including coconuts and papayas that climbs a long ridge forming the crocodile's back to the peak of the island, 463 metres high. It then dips gently down in the north to form a distant snout looking out into the wide Pacific. Far beyond the horizon, beyond Tonga, Tahiti and the Cook Islands, lies South America.

It looks idyllic, but Rabi has a painful history of migration from far across the Pacific with a struggle to maintain uprooted identities and traditions.

This same view across the water met Iotua's grandparents in 1945 when the British colonial authorities shipped more than 1,000 people here to Fiji from their ancestral home on tiny Banaba island in the central Pacific, about 2,000km away to the north-west and now part of a different island nation, Kiribati. Banaba had been ruined by British mining of phosphate, a natural crop fertiliser formed from thousands of years of bird droppings, and ravaged by Japanese occupation during the Second World War.

At the end of colonial rule in the 1970s, Banaba became part of the Pacific island nation of Kiribati and Rabi (an island in newly independent Fiji), leaving families shipped to Rabi from Banaba and their descendants feeling torn between the two nations.

The almost forgotten story[1] of the 1945 relocation and the Banabans' struggle to preserve their culture straddling two Pacific countries are now suddenly again relevant for low-lying island nations worried by a completely different threat in the twenty-first century: sea level rise that may force migrations. In a sign of how far Rabi is rooted in two cultures, Rabi islanders even get to select members of national parliaments in their new home in Fiji and their ancestral home in Kiribati.

Many Pacific nations are extremely vulnerable to sea level rise and international law has few pointers about how to deal with a possible surge in migration caused by climate change. The Banabans' forced migration to Rabi is an imperfect blueprint, but the settlers and their descendants have already tackled legal and emotional pitfalls linked to identity, right of settlement, passports, land ownership and voting.

The Banabans were brought here, exhausted, in cyclone season on 15 December 1945. With scant supplies, they were left, living in tents, to rebuild their lives in a new country. They had to overcome years of hardships. Iotua is returning to prepare for anniversary celebrations of the resilience of the Banabans after their arrival. The Banabans' successes and failures with relocation may help other governments plan for a more humane future if islands become uninhabitable because of sea level rise.

Rabi is a far cry from Banaba, a crescent-shaped rocky island covering about 6 sq km. By contrast, Rabi island is 67 sq km and has huge areas for agriculture, a shift from dependence on mining, fish and coconuts in Banaba. Rabi is exploding with vegetation all the way along its crocodile shape.

Many of the people now living on Rabi and their descendants, like Iotua, feel mainly Fijian. That's often expressed viscerally as "blood". "We want to call ourselves Fijian, but we don't have the blood," says Iotua, who was born in Suva, Fiji's capital, to a father from Rabi and a mother from Kiribati. She has worked for the office representing Rabi's interests in Suva. Mostly, she says, she just feels "Rabian".

It is a gruelling island-hopping trip for her to get here from the capital – an overnight ferry, then three hours bouncing along in a truck with more than twenty other people crammed in the back

– including me – with piles of luggage on a largely unpaved road. Finally, there is the forty-minute trip by a small launch to Rabi, requiring wading knee-deep into the sea to clamber aboard.

Once in Rabi, the passengers again wade, past an ageing concrete sea wall that forms the port and a sign that warns, "Be careful. Falling coconuts", and head off in trucks or on foot to their family homes. A white Methodist church is the main landmark above the main village. They wave to children playing rugby and running around on a pitch where the settlers first landed in 1945. On one beach nearby along the coast, a pig is grunting, wallowing contentedly in the shallows, its snout just above the water and its neck tethered loosely to a rope dangling from a tree.

Like Iotua and many others, Tewaite Tewai, who is in her fifties, also feels caught between cultures. Her mother was one of the first settlers from Banaba. "I have the blood of a Banaban. But we're living in Fiji. We tell Fijian jokes, we eat what Fijians eat. We live like Fijians," she said around a kitchen table of the local guest house. "I'd like to visit my mother's home but not to live there," she said of Banaba, also known as Ocean Island after the British merchant ship, *Ocean*, which was among the first European vessels to sight the island in 1804.

That wish to visit is a common refrain and stops short of wanting to return to live on Banaba, now home to about 300 people. Few people on Rabi have ever been to Banaba, one of the hardest places to reach in the Pacific, where ships rarely call since the phosphate mines closed in 1979.[2] Almost all of the first generation of settlers on Rabi have passed away.

"My heart is still in Ocean Island," one of those survivors, Tonganariki Taratai, who was 4 years old when she arrived in Rabi from Banaba in 1945, told the *Fiji Times* in 2015. She said that she and others had managed to maintain their culture despite being uprooted across the Pacific.[3]

Tewai noted that Banaba, a tiny drought-prone island, became desolate and steadily more dependent on imports of food after coconut palms and other food plants were bulldozed to make way for phosphate mines. She cracks into laughter when she tells me one children's song on Banaba – lightening hardship into a melody: "What time will the flour come, when will the sugar come? There's nothing in the shop."

In normal times – like when I visited in December 2018, before coronavirus and cyclones battered the economy – Rabi is vibrant. The island economy depends heavily on kava, the pepper plant whose ground-up roots are used to make a drink served from a coconut bowl. Rabi also has coconut plantations producing copra, the white kernels that are the source of coconut oil.

The 1945 arrival in Rabi lives on vividly in stories of stubborn survival. At the end of 1945, 1,003 people from Banaba were collected up aboard the *Triona* ship and sent to Fiji, in the first and biggest wave of migration to Rabi. They consented, under pressure, to move.

British authorities bought Rabi with money owed to the Banabans after decades of mining phosphate rock scraped away the topsoil of their island. When Kiribati gained independence from Britain in 1979, Banaba became part of Kiribati.

The arrival in December 1945 was a massive culture shock. Rabi is in the tropics but felt chilly to the new arrivals. It was covered with jungle and wet compared to Banaba, which is almost on the equator and more prone to droughts than downpours.

David Christopher – a Rabi leader who is an athletic man in his early seventies, about 190cm tall and with an easy charisma – said the failure to visit Rabi in advance was a massive error. But he said the Banabans were given little choice by the British colonial authorities. It's a lesson for any future migrants displaced by sea level rise to check their new homeland is suitable.

Before embarking, the Banabans were even shown photographs of fine-looking Fijian houses but were not told that these buildings were in the capital, Suva, not on Rabi. On the island there were no homes, no clinic, no shops, no roads. They had to live in tents. Forty of the elderly died in the first few months after arrival – measles, lung disease and diarrhoea were frequent in the wet climate.

"The basic thing is that before you move people should have a look ... to feel for themselves, to see for themselves if it is suitable, to their liking, and if the basic amenities are there," Christopher said. "That did not happen. What happened was that we were told a different story. It was the end of the war, people were tired, they were sick. They were told: 'Go to Fiji, there are big houses there, how wonderful.' And they came here and there were no houses."

"There were two factors: they were told lies, as it were, and the other factor was they did not have the urge to argue" after suffering during the war when Banaba was occupied by the Japanese, he explained, "Maybe they should have said 'wait, we want to see first, maybe you are not telling the truth'. But they did not."

The new arrivals landed on what was to become the sports and parade ground, which now has rugby posts for Fiji's national sport. "It was very much a refugee camp, or maybe worse," Christopher said. The first night the Banabans put up tents on the level grassy field by the waterfront, thinking they would at least be able to rest after the voyage. But then in the night came a new shock, with what the settlers thought were "monsters" attacking in the dark.

"There were a lot of cattle here in those days ... In the daytime the cattle went up the hill and in the night they came down," Christopher said. "The cattle come down, they don't know there are tents, they walk through the tents and the people they yell and cry," he said. "They don't know what these strange animals were." Banaba has no large land animals – it is a refuge for seabirds, and crabs are among the few creatures scuttling on the ground.

The Banabans were also alarmed by the bubbling of a small creek nearby flowing into the sea. "It was a big river to them, there are none on Kiribati. They were frightened by the cattle, the mountains, the rocks, the creek," he said.

That shock felt by the original settlers has been numbed by the intervening decades: several people on Rabi laughed when they told me about the shock of the "monster cows". "We feel sorry for them or we laugh at them – I don't know," Christopher said. It's also a story of survival against the odds, the resilience of the human spirit and of the ability to adapt to a new environment.

The settlers named the four villages in Rabi after those on Banaba: Buakonikai, Tabwewa, Tabiang and Uma. But as the cows and the river show, there's no place like home – even if it's been ruined.

The settlers also had problems figuring out which plants were edible. Early settlers who hacked their way into the jungle to explore the island also upset swarms of hornets and ended up being attacked by the aggressive, stinging insects, unknown to those from Banaba.

Many islanders say the obvious – you can't just load people onto a ship and dump them in a new nation without resolving basic issues like food, water, housing, jobs or the right to stay. And the settlement shows how complicated moving entire populations would be if oceans rise fast, even to a virtually uninhabited island.

Since those first arrivals, Rabi has adapted to an existence split between two nations. Exemplified by Christopher, Rabi now has one of the most extraordinary parliamentary systems in the world, with representation in both the parliaments in Fiji and Kiribati, 2,000km away across the Pacific. The rights of Rabi islanders to sit in the Kiribati parliament were described in the British parliament in 1979, the end of the colonial period, as "unique provisions"[4] in the entire empire.

At first glance, the Rabi islanders' seat in Kiribati's parliament seems like allowing a Greek citizen to vote in the British House of Commons, a Mexican to debate in the US Senate or a Russian to sit in the Japanese Diet. But in the history of the island it makes sense. Recognising the islanders' roots in Kiribati, Kiribati allows the Rabi Council of Leaders to nominate – rather than allow everyone to vote for – a representative for the forty-five-seat House of Assembly, in Kiribati. The Rabi islanders separately vote for a member to the Fijian parliament from an eastern constituency which includes Rabi.

Christopher, famed on Rabi for his booming voice and infectious laugh, was a member of the Fijian parliament from 2001 to 2005 and then nominated to the Kiribati parliament from 2014 to 2019. "I'm a Fijian citizen, I have a Fijian passport, I travel there [to Kiribati] on my Fijian passport. I attend parliamentary meetings overseas from Kiribati with my Fijian passport," Christopher told me when I visited him in an office situated in a large wooden building on Rabi, emblazoned the 'Rabi Council of Leaders'. "I could have a Kiribati passport, but I don't," he said, as the rain pattered on the corrugated roof, turning the road outside into a deep, brown mud. He said that sometimes raises eyebrows at passport checks at airports abroad when he is in a delegation with Kiribati parliamentarians and his light blue Fijian passport stands out from the darker blue of his colleagues'.

People living on Banaba island also elect a member of the Kiribati parliament. The right of Rabi islanders, far away in Fiji, to select members of parliament in Kiribati is sometimes criticised by other politicians in Kiribati as granting too much power to a minority living abroad.

Still, Rabi islanders say that their right to choose politicians in both Fiji and Kiribati reflects a fraught history straddling two nations. Would any future migrants, forced to move by rising seas, retain parliamentary seats in their former homelands and get new ones too?

Kiribati is especially vulnerable to sea level rise. The nation, of 110,000 people, comprises thirty-three coral atolls and islands, many a few metres at most above the water. It also covers a vast area – it is the only country with territory in each of the world's four hemispheres, with islands both north and south of the equator and east and west of the international date line.

Ironically, given how much of the island has been blasted away by mining, Banaba is now the highest point in Kiribati at 87 metres, high enough to survive any amount of sea level rise caused by melting ice from Greenland to Antarctica. Some in Kiribati imagine it could be a bastion to keep the national flag flying, asserting continued existence even as other islands vanish beneath the waves.

Fiji's 900,000 people are not so vulnerable because most live on large, volcanic islands rather than fragile coral atolls.

Christopher welcomes me to one meeting at which we sit cross-legged on the floor drinking kava from a coconut shell, sipped then handed around to others in the room. It has a pleasant peppery taste, turning my tongue and lips a bit numb. He says most people on Rabi island feel ever more Fijian as the decades pass. Many Banabans have moved to other parts of Fiji, meaning there are probably 10,000 people of Banaban descent in total in the nation.

"I am born here, I am raised here, I did all my boarding school here ... I am very much more Fijian than Banaban, as compared with my parents, who came direct from Banaba after the war," Christopher explains. "I am used to the weather, the food, the climate, the language. I speak Fijian, English and Gilbertese [spoken in Kiribati] – three languages. That keeps me going," he said with a laugh.

Under the Rabi model, Fiji has been accommodating – anyone born on Rabi to Banaban ancestors now is automatically Fijian, and Fijian laws apply to them. Fiji granted citizenship to all Banabans and descendants living on Rabi in 2005.[5] That integration of the community into Fiji defuses many of the worrying issues of climate migrants: do you need visas, work permits? Can you set up an enclave in a new country and apply laws of your home state, and vote in an election abroad? Can you get rights to stay permanently, or will you risk being expelled, for instance, if your new host government turns nationalistic or xenophobic?

For Christopher, just getting to work in Kiribati to represent Rabi in parliament is an odyssey, even before travel shutdowns caused by coronavirus in 2020. In normal times, his two- to three-day commute – taken three times a year when the Kiribati parliament sits – starts with a stroll to the dock. He then takes the motorboat over Vanua Levu island, often getting drenched during the trip. (When I took that same trip leaving Rabi, the boatman had to stop to bail out water and the passengers had to hold up a big piece of plywood in the bow as a makeshift shield to help stop the waves swamping the boat. It worked.)

From the landing spot, Christopher catches a bus, a truck or rides a taxi to either Savusavu or Labasa, which have airstrips. From there it's an hour by a small Twin Otter propellor plane to Nadi – Fiji's international airport on Viti Levu, the biggest island. He has to stay overnight in Nadi because the three-hour flight to Tarawa, the capital of Kiribati, always leaves early in the morning. Finally, after the exhausting trip to Tarawa, it's a taxi ride to parliament.

Some parliamentarians elsewhere travel further to represent their people – it's about 5,400km from Anchorage, Alaska, to Washington DC, and close to 8,000km from the easternmost part of Russia to Moscow. Perhaps most extreme, New Caledonia is part of France and elects deputies to the French National Assembly and the European Parliament, 16,500km away on the other side of the globe.

Unlike them, however, Christopher has represented people in the parliaments of two different nations. He reflects that the style of debate is very different in each parliament: "In the Fijian parliament, you cannot fall asleep in chambers, there's too many rowdy and noisy interjections." By contrast, Kiribati is more sedate: "There are no interjections except with the approval of the speaker."

He's a widower with two children who live in Kiribati – he stays with them when he goes to work. He also has a daughter in New Zealand. "I vote with the government of the day," he said of the Kiribati parliament. "We have been advised by our elders that 'you go there, you have plenty more to gain if you join the government than if you are in the opposition'. I think that makes some sense."

Rabi's legal umbilical cord to Kiribati lets it apply for some funds, especially to help get scholarships for students studying in Fiji and other nations. As citizens of Fiji, the inhabitants of Rabi also have rights to rural development funds from Fiji. But Christopher says there are limits.

"We don't want to be doing double dipping, because we are looked after by the Fiji government here, our basic needs are met here," he said. "But anything over and above, if we can grab from Kiribati, why not?" he adds, bursting into laughter.

He shrugs off one potential hurdle – like in many nations, parliamentarians in Kiribati have to swear an oath of allegiance. Technically, standing and pledging allegiance to Kiribati could be seen as a reason by Fiji to revoke his citizenship. "I do [take the oath], it's OK. I haven't got chased out of Fiji yet," he jokes.

In Kiribati's parliament, Banaba's representative is among the most active, seeking funds, including for a ramp up the island from a port – an idea alien to low-lying islands in Kiribati. In one inconclusive exchange[6] from 2020, Banaban representative Tibanga Taratai asked: "During the last four years, this Government said it would construct a new ramp for Banaba. I want to ask this Government to explain the status of this project?" A minister replies: "The concept, design and costing have been completed and funding is currently sought."

In 2021, Rabi's economy is struggling, a sign that the relocation so many decades ago has not secured prosperity. After decades of exploitation, the original Banaban workforce knew more about hacking phosphate rock from the ground than tending crops.

After two cyclones in December 2020 and January 2021, "some homes were completely damaged while many needed ... general repair and maintenance," Christopher said. "All our breadfruit trees – which were bearing at that time – were completely wiped out, leaving only the bare trunks standing. Coconut trees were uprooted or broken down. Those left standing are without fruit and some leaves only. Mango, lemon, tapioca, pawpaw and other such plants grown in our back yards were also gone," he told me in an email.

Staff working for the Rabi Council, which manages the island, have cut hours to three days a week, while the Banaban Virgin Coconut Oil factory has stopped production after a fall in tourism and demand from Fijian holiday resorts, he added. And the price of kava, which used to be between 100 and 120 Fijian dollars (US$50–$60) per kilo, has halved to about 50–60 Fijian dollars, he said, forcing some farmers to revert to cutting copra.

The history of the Banabans' woes began in 1900 in an office in Sydney, Australia, where Albert Ellis,[7] a young New Zealander, noticed a strange whitish rock used as a doorstop. He discovered that the block, which had come from the island of Nauru near Banaba, was almost pure phosphate, a valued fertiliser. He travelled to Banaba, then home to about 450 people, and realised that much of the island is high-grade phosphate.

The Banabans, with little or no understanding of legal contracts, signed away mining rights for what was a derisory £50 (sterling) a year for 999 years, many signing the contracts with an 'X' since they could not write. A good number of the islanders turned to helping the mining operation. They did backbreaking work, pushing wheelbarrows full of phosphate up makeshift ramps into British ships. Britain then annexed Banaba. Later, the Banabans secured a small share of royalties but became steadily more disenchanted with the mining that was ruining their island, stripping away the land and kicking up clouds of dust that choked their children.

Eventually, the Banabans ended up as a minority on their own island as the British Phosphate Commissioners (BPC) – a mining venture run by Britain, Australia and New Zealand – recruited workers from other small Pacific islands, as well as from Japan and China.

McAdam, the law professor at the University of New South Wales, said British colonial authorities were clearly aware that mining would make Banaba uninhabitable. Official UK documents from the early twentieth century even noted it would be "repugnant" to "evict a native tribe" simply "to afford wider opportunity of gain to a rich commercial corporation". But she wrote that the colonial authorities concluded that the benefits of fertilisers to grow vital crops in Australia and other nations outweighed the suffering of a few hundred Banabans.[8] "The interests of the Empire seem to demand that the process of development on Ocean Island should be allowed to continue until the whole island is worked out," Britain's High Commissioner for the Western Pacific, C.H. Rodwell, wrote bluntly in 1919.

A 1977 BBC documentary, *Go Tell It to the Judge*[9] about Banaba, tracked the destruction of the island. At the time, it caused widespread shock in Britain about the mistreatment of the Banabans. It shows how in 1928, Arthur Grimble, resident Commissioner of the Gilbert and Ellice Islands, a group that included Banaba, called a meeting to try to persuade the Banabans to sell more land, despite opposition led by Banaban leader Rotan Tito. Grimble starts off by telling islanders including Tito, "The BPC provides phosphate to the farmers of Great Britain, of Australia, and of New Zealand, to the farmers of the British Empire. You, your island, is one of the only two places in the whole British Empire producing this phosphate. You should be very proud."

When Tito objects that the money offered is far below the market price, Grimble retorts, "You are thinking selfishly. You are not thinking of your children and of their future. If you sell no more land to the BPC, you are signing your own death warrant. Your people will die. You will be committing suicide."

Later, when a group of women similarly object ferociously to the offer to buy their land, Grimble tells them that they will need the money to pay for food and doctors. "The land lying idle will grow only coconuts. You are no longer used to living only on coconuts. Your food now includes biscuits, it includes rice," he tells them. When they shake their heads, he asks, "Do you realise how foolish, how utterly foolish your attitude is?"

Afterwards, he wrote, "Talking to them, one had the hopeless certainty that they were not listening, simply sitting and repeating to themselves, 'we want to keep our land, we want to keep our land', and so on ad infinitum." He said that he explained that "I was giving conscientious advice, as a friend". Grimble also threatened them with unspecified "punishment" for insulting King George V.

The BPC eventually obtained a compulsory purchase order for the land, sidelining the Banabans for what the British viewed as the greater good of the empire. And that was how it ended, despite the growing disenchantment by the Banabans. Women sometimes clung to the trunks of coconut trees set to be cut down: they were forced aside to make way for bulldozers.[10] When mining on Banaba halted in 1979, 21 million tonnes of phosphate had been stripped from Banaba – almost equivalent to the weight of four Pyramids of Giza.

In a debate in the UK parliament that same year, Conservative Bernard Braine[11] said the Banabans were paid just 11¾ old pennies (5 pence) a ton for phosphate rock, less than a tenth of the sales price in Australia. "In moral terms, this amounted to fraud perpetrated with the assistance of blackmail," he said.

Even by the 1940s the island was so degraded by mining that the Banabans reluctantly embraced the idea of a move to Fiji, something that was legally far easier than now since both were under British colonial rule. Any mass migration in the twenty-first century, driven by climate change, is likely to be far more complicated across international borders.

For the Banabans, their homeland had become an alien wasteland, scoured by phosphate excavations and housing for miners and their families and a dock for loading the rocks. There was even a rocky nine-hole golf course for the privileged expats. "The balls had to be painted yellow or red so they stood out against the background of white coral," Katerina Martina Teaiwa wrote in her book, *Consuming Ocean Island*.[12]

Teaiwa also once visited the rusting remains of abandoned machinery on the island in the 1990s. "It was like looking at a post-apocalyptic zone," she said. A land that had belonged to the Banabans – their source of traditions, dances, song and food – had been destroyed.

History has many examples of the suffering caused by forced relocations, from Spain's decision in 1492 to expel Jews to China's relocation of more than a million people to build its Three Gorges Dam that opened in 2003. One study in 2005[13] found eighty-six examples of communities moved in the Pacific, mostly locally by villages within nations, often because of natural disasters such as flooding or cyclones.

The Banabans' relocation to Rabi features among these Pacific uprootings, alongside 167 people from Bikini Atoll in the Marshall Islands in 1946, who were forced away to clear the island for US nuclear bomb testing. Near Rabi is Kioa island, bought by settlers from Vaitupu island in Tuvalu after the Second World War. Those settlers now have a similar legal status in Fiji to those from Rabi. The remoteness of the islands, and the lack of crowding in Fiji, helps limit frictions with local Fijians.

In the early 1940s, Rabi was put on the market by the Lever Brothers Pacific Plantations Proprietary Limited for £25,000. The plantations were producing copra. Rabi was also attractive for the British as a potential home for the Banabans partly because only a few people, including from the Solomon Islands, lived on the island in the 1940s. That meant no mass unpopular displacement of Fijians.

But the Second World War broke out before any Banabans could travel, as originally planned, to inspect the island to let them make an informed decision about relocation. And then the Japanese occupied Banaba for three years from 1942. A third of the Banaban population, or about 400 people, died during the Japanese occupation, "some from starvation, others were poisoned, shot, beheaded or died from electrocution," wrote Julia B. Edwards, a researcher on climate change and resettlement at the Pacific Conference of Churches in Suva, Fiji. Many others were sent to internment camps on other islands.

During the war, Britain went ahead and bought Rabi in 1942. After the war, surviving Banabans were told their home island was no longer habitable, leaving them no choice except to travel to Rabi. Despite their exhaustion, the Banabans resisted.

"It will take some years for the Banaban community to recover from their treatment during the Japanese occupation: they were only a shadow of their former selves when discovered by the allied occupation forces," wrote H.E. Maude, a colonial administrator. He also noted that the Banabans were more openly sceptical of Britain after the war. "While for years they have distrusted the government's good faith, they are now said to be more openly critical than before, which is ascribed to their having seen the European beaten, if only for a while, by a brown-skinned race such as themselves," he wrote in a report.[14]

Among issues discussed in the 1940s were how far the Banabans would live in a sort of enclave, with their own laws and customs, or be free to travel and integrate into Fijian society. Initially they were not allowed to travel to other parts of Fiji. But even colonial officials wrote at the time that it would be hard to restrict Banabans to Rabi as if they were in prison or like "the lepers on Makogai". Makogai was a Fijian island used to isolate the sick.

McAdam, the Australian academic, quotes Harold Cooper, the chief of the Government Information Office for the Fiji Islands, as writing around 1946: "They cannot remain forever a separate community, wrapped in a sort of geographical cellophane, and if they are to become happy and useful inhabitants of this Colony they must get to know, and to understand, the people who are to be their neighbours." McAdam also notes counter-currents likely to be faced in future: What guarantees do the new arrivals have that they can stay, or have rights to fisheries offshore that may conflict with those of neighbouring islands?

Some Fijians were sceptical of the settlers. One author in the *Fiji Times and Herald* newspaper, wary that the Banabans might undermine jobs for Fijians, sniffed in 1946, "If the Fijian people as a whole could be consulted they might say they do not want the Banabans here at all."

The original idea was always that Banabans could go home if they wanted after a trial period. But it was an empty hope, since virtually everyone knew Banaba was wrecked. Banabans on Rabi, in a secret ballot in 1947, voted by 270 to 48 to stay, rather than return to what had been their home.

In the end, the Rabi Council was granted a large amount of autonomy to manage the island's affairs, such as healthcare, schools and rules for land inheritance. It still has greater powers to raise taxes and make local regulations than other islands within Fiji. But Fiji is still in charge. The Rabi Council, the main employer on Rabi, has suffered from corruption and mismanagement down the years – it has sometimes been placed under administration by the Fijian authorities.

The Banabans technically 'owned' the island after they arrived, but a 2013 Fijian constitution weakened Banaban land rights on the island by introducing the idea that it could be compulsorily bought for "public purpose". "It essentially makes the Banabans tenants-at-will of the [Fijian] state," McAdam wrote.

When Fiji became independent from Britain in 1970, most Banabans chose to become Fijian, rather than retain British colonial citizenship. Fiji gave citizenship automatically to all Banabans who had been born in Fiji, or about 1,500 of 2,000 people then on Rabi island. But it stopped short of granting citizenship to those who had been born in Banaba.

"Fiji refused to confer automatic citizenship on all Banabans, because it would have meant treating them differently from other minority groups in the country," McAdam wrote in the *Law and History Review* in 2016. It could, for instance, be seen as discriminating against Fiji's large Indian minority, brought in as labourers for sugar cane plantations, who got no automatic right to citizenship. Many of the older Banabans, however, did not seek Fijian citizenship, preferring to remain British.

In the mid-1970s, the Banabans launched a high-profile bid for more than £20 million in compensation from Britain in the London courts. They alleged that the British had failed to restore vegetation in mined land on Banaba and that they were not paid a fair price. In a 1977 ruling after a case that lasted more than 220 days, they lost. Judge Robert Megarry, who had even travelled to Banaba to inspect the island, concluded that the existing laws did not allow him to make an award just because the Banabans "have had a raw deal". He left it to the British authorities to make an award. Eventually, Britain, Australia and New Zealand offered Australian $10 million[15] to the Banabans, about a third of their original claim, and with the condition that the Banabans took no further legal action. The Banabans eventually agreed in 1981.

The Banabans sought full independence for Banaba as a nation, a drive that also failed. In 2005, Fiji granted citizenship to later waves of Banabans who arrived after Fijian independence, plastering over another potential source of conflict. In Kiribati, Banabans have rights of entry and to acquire citizenship. Unlike other nationalities, for instance, Fijian passport holders from Rabi can arrive on a one-way plane ticket, rather than need a flight booked out.

Forty years after accepting a settlement from London, Rabi sees farming as the future of the island. "We believe our future is in agricultural development," Christopher said. However, as previously noted, prices of kava have been unpredictable. They spiked after Cyclone Winston in 2016, which killed forty-four people in Fiji and destroyed kava production in other parts of the nation. That brief spike helped Rabi's economy – some islanders invested their extra kava earnings in buying bicycles, extending homes and buying new furniture.

Christopher would like to see a new road built across the island to open up the eastern side to more economic activity – so far the main, unpaved road goes along the flatter coast on the western side: "There is plenty of land, and good land, but the problem is that it's not accessible. Most of the plantations are on the other side of the island. There is no road, we go by boat, or you walk or you go on horseback."

Rabi also wants to help those still on Banaba.

Banaba, Christopher says, could open up to tourism in future, especially for descendants of those who once lived there. That would mean both Banabans in exile wanting to see their ancestral land and the descendants of phosphate miners – such as workers from China and Tuvalu – or of Japanese soldiers who died during the war. In recent years, only one ship a month visits Banaba, from Tarawa, the capital of Kiribati, and the schedule is unpredictable. On one of his visits in the early 1980s, Christopher got stuck, "We were supposed to be there a month, but the boat didn't come for three months. We survived, but I don't want to get stranded there again."

As seas rise, Rabi is also vulnerable to coastal erosion, even though the island's hard volcanic base is less at risk than corals. "We built some walls but it would be difficult to build a wall around the whole island," Christopher says. In some places, the walls have washed away, exposing homes along the coast to flooding.

With rising seas putting many low-lying nations at risk from the Pacific to the Caribbean, many islanders are reluctant to talk about a future when people may have to abandon homes because of climate change and rising seas. Many islanders have rightly viewed climate change as the fault of industrialised nations and want to raise pressure on major emitters, such as China, the US and the European Union, to make cuts in emissions to contain the problem, rather than consider moving.

In a speech to environment ministers at a UN conference in Poland in 2008, Tuvalu's then Prime Minister Apisai Ielemia stated the case bluntly: "It is our belief that Tuvalu, as a nation, has a right to exist forever. It is our basic human right. We are not contemplating migration. We are a proud nation of people with a unique culture which cannot be relocated somewhere else. We want to survive as a people and as a nation. We will survive. It is our fundamental right."[16]

But, with seas lapping higher, moving may be inevitable.

The resettlement of the Banabans gives some hints to whether it will be possible for islanders to retain their culture, sense of identity, social networks, family and history in a new nation with different climate, geography, foods, plants, fish, birds and animals – perhaps including the odd herd of cattle.

Such issues have hardly been considered so far in the international debates about climate change, but Fiji is on the leading edge. The government, which hosted UN climate negotiations in 2017 in Bonn, has said it is willing to do its share to take in migrants from other Pacific islands if needed.

"Rabi island was used as a sanctuary" for the Banabans, Fiji's Attorney General Sayed-Khaiyum told me. "At the time it was a special case," after the Second World War. "But this time is a special case too," he says because in many parts of the world climate change is affecting coasts. He says Fiji would do its share to reassure other Pacific nations: "We don't have all the answers. There is more of a psychological thing that, at the end of the day, you won't get abandoned. We can't see our Pacific islanders go under water." Fiji has an office to look into the legal aspects of migration such as visas, rights to employment permits and citizenship.

Most migration experts say that people are likely to move in small numbers, rather than mass relocations as happened to the Banabans.

As previously mentioned, Kiribati's government bought 6,000 acres (2,400 hectares) of forested land in Fiji in 2014 to help secure food production for the low-lying Pacific nation. As seas rise, king tides and storms can blow salt onto farmland, blighting crops. In the longer term, the land could provide a second home for citizens from Kiribati, if Fiji agrees.

"With sea level rise, the value of land will go up," former Kiribati President Tong told me in 2014. He also pioneered the idea of "migration with dignity", to avoid the shock of forced, mass movement that would turn proud islanders into refugees. Under his concept, more people from Kiribati might study or work in places like Sydney or Los Angeles, and then return home to educate others about the opportunities and risks. Gradually, that could seed communities living abroad – where those who have already emigrated could guide new arrivals through the maze of applying for residence, jobs and other rights.[17]

But new arrivals need to know they can stay.

"Land tenure is a critical factor in relocation," according to a study of Pacific relocations in 2005 led by John R. Campbell at New Zealand's University of Waikato. "Where communities can relocate within their own territory, friction and tension can be avoided much more easily. Any movement beyond a community's boundaries is likely to require a high level of consultation and negotiation with the 'host' community," the study states.[18]

However, it is almost impossible to guarantee land rights forever, since that means carving out a foreign enclave. A less permanent option would be for the host nation to lease land to immigrants – but that leaves a lot of uncertainty, and risks of expulsion if there is a nationalist backlash.

Defining land ownership is often fraught with difficulty. Around the world, even embassies or military bases run by foreign powers are still formally part of the territory of the host country. The US naval base at Guantanamo Bay in Cuba, for instance, originates from a lease signed by both nations in 1903 that Havana has viewed as illegal since its 1959 communist revolution.[19]

McAdam notes, "There is much more to relocation than simply securing territory [...] Those who move need to know that they can remain and re-enter the country, enjoy work rights and access to health care, have access to social security if necessary, be able to maintain their culture and traditions, and that the legal status of children born there is clear."

Over the years, there have been talks about relocating entire populations. In 2010, the African Union said it would look into offering land to Haitians after their country was devastated by an earthquake. The idea had echoes of the creation of the nation of Liberia in West Africa in the nineteenth century by freed American slaves.

Closer to Fiji, the island state of Nauru in the Pacific, the world's smallest republic, was rich in phosphates and at risk of being mined out of existence by the early 1960s. Finding a new home, following the example of Rabi, was an obvious possibility for the 4,600 population. Australia, Britain and New Zealand – who had power over Nauru under a UN mandate – agreed in 1960 that the best option was the gradual resettlement of Nauruans in their three countries, over thirty years, according to one study.[20]

The offer included: "(1) citizenship; (2) equal opportunity and freedom of social contact; (3) for young people, education to the fullest extent of their capabilities and an allowance of £600 per year for five years, after which they would be assisted to look for suitable employment; and (4) for adults, employment in any of the three countries, passage, a house, maintenance for a six-week period, further training for self-employment, and eligibility for all social welfare benefits," the authors wrote.

But there was an immediate opposition to granting territory to any settlers. Australian Prime Minister Bob Menzies wrote in a letter in January 1962 that "no Australian Government would be likely to agree to the establishment of a separate Nauruan community as an enclave within the borders of Australia". Australia eventually offered Nauru the possibility of moving to remote Curtis Island, off the coast of Queensland, with limited self-rule. Curtis Island is more than thirty times bigger than tiny Nauru, which is just 21 sq km.

But Nauru said in an official statement in 1964 that the offer would mean a loss of identity and also feared racism within Australia, whose laws at the time explicitly favoured white immigrants. "Your terms insisted on our becoming Australians with all that citizenship entails, whereas we wish to remain as a Nauruan people in the fullest sense of the term even if we were resettled on Curtis Island. To owe allegiance to ourselves does not mean that we are coming to your shores to do you harm," they wrote.

Resettlement negotiations collapsed in 1964. Nauru won independence in 1968 and, with it, full control over phosphate resources which meant a boom that briefly gave its citizens one of the highest per capita incomes in the world. But Nauru's boom led to massive over-spending, mismanagement and bust.

As Banaba shows, there are many perils for Pacific islanders trying to weigh the benefits of relocation or staying as seas rise.

On a visit to Fiji, as part of a trip to the Pacific in 2019, UN Secretary-General António Guterres said the world should listen to Pacific islanders on the frontlines of climate change: "As coastal areas or degraded inland areas become uninhabitable, people will seek safety and better lives elsewhere. In 2016, more than 24 million people in 118 countries and territories were displaced by natural disasters — three times as many as were displaced by conflict."

"Your experiences underscore the urgency of the threat. The Pacific has a unique moral authority to speak out. It is time for the world to listen," he said.

6

IN THE CARIBBEAN, THE DREAM OF HIGHER GROUND

The sun gets closer every day.

– a saying by inhabitants of the Caribbean island of Gardi Sugdub off Panama

A planned move of the Guna Indigenous people from an overcrowded Caribbean island off Panama was intended as a showcase for Latin American nations struggling with sea level rise. But more than ten years later the pioneering move has stalled – a hospital and school on the mainland have been partly built, then abandoned to bats, spiders and lizards. The islanders now hope to move to new homes in 2023.

PABLO PRECIADO WANTS TO MOVE to the mainland from his low-lying coral island home off Panama, fed up with seeing sardines swimming in his house when storms whip up the blue Caribbean into an angry grey flood. It's already been a long wait.

A local leader of the Guna Indigenous people on the tiny coral island of Gardi Sugdub, Preciado, a soft-spoken man in his early seventies, has long helped guide a plan to relocate about 1,450 people to the mainland to escape rising seas and overcrowding.

Agreed upon by the Guna people in 2010,[1] the wrenching move has been repeatedly delayed by a lack of Panamanian government and other funds along with a litany of mismanagement, Byzantine corruption scandals and disputes over drug smuggling. And, most recently, the coronavirus pandemic. The Guna now expect the move to the mainland in early 2023.

All sides agree urgent action is needed for Guna Yala province, a string of about 365 islands including Gardi Sugdub – one for each day of the year, as the locals say – off a ribbon of densely forested land about 200km long on the Caribbean coast stretching to the Colombian border. About 33,000 people live in the province.

The Panamanian government stated in a December 2020 report to the United Nations, outlining its plans for combatting climate change in the coming decade, that Guna Yala marks a "critical case" of sea level rise for the Central American nation.

So far, however, the attempt to move the population from Gardi Sugdub to the mainland, about 1.5km away across the turquoise bay, is more of a cautionary tale than a model for relocation in Latin America as seas rise. Millions of dollars were spent to build a school and health centre on the mainland coast near Gardi Sugdub half a decade ago, including funds for the school from the Inter-American Development Bank. But the projects stalled and plants and animals including bats and lizards moved in. Air conditioners were bolted to some walls, but the plugs were left dangling as there was no electricity. In 2021, 300 houses for the Guna are finally under construction on a plot of orange-brown soil, after a pause caused by the coronavirus, in a US$9 million housing project led by the Panamanian government.

Preciado, who has been the local political and spiritual leader or 'Saila' on Gardi Sugdub for several of the past years, has long been philosophical about repeated delays in moving from the sandy island where the highest point is less than a metre above sea level. "We don't want to move until everything is in order," he told me when we met in late 2018, prescient about coming delays and tinged by the disappointment of the long wait.

He lives in a house in the heart of the island, with walls made of canes built to sway in the wind, rather than snap in storms. They are neatly tied together with plastic thread, beneath a roof of palm leaves woven tight together to keep out frequent downpours.

Yet even here, on a frontline of climate change, he says that some people aren't convinced that the sea is rising, ignoring satellite and tidal gauges showing that the level of the Caribbean is creeping higher as ice sheets melt and glaciers pour more water into the oceans. Part of the misunderstanding is because the Guna have built protective walls against storm surges in recent decades that have both expanded and raised the island, he says. This strategy, of adding corals, rocks and sand, has enabled the rising level of the island to exceed gains in sea level in some places. On the downside, the reinforcement of the islands is damaging the reefs that are nurseries for fish.

To illustrate the trends, Preciado gingerly stands up, puts his hand down to touch his knee and says, "The water used to come up to here" during storm surges in the 1970s and 1980s when the harshest winds sweep across the Caribbean late in the year and drive the sea level higher. "In the worst case there were sardines swimming here," he said, smiling and waggling his hands to imitate the movement of fish streaming through his house. "But now we're filling up the islands with rocks so the water doesn't come so high." He and other residents say the waters typically now come up only to their ankles when tropical storms strike because of the filling of the islands.

If you get fish swimming in your house, he advises, there's little you can do. Eventually the grey waters and the debris subside, as the tide goes out or the winds ease. "You just have to wait. It can take 1, 2 or 3 hours, just waiting for the water to decline."

But the fish drive home a serious point. A few local sceptics about climate change who don't want to move – mostly elderly people – see the lack of an obvious rise in sea level as an argument that nothing is happening. "Some people say it's all lies. 'How can an island sink?'" Preciado asks.

But he's well versed in the risks of global warming. "Scientists realise that the big glaciers are breaking up. They say that these islands won't exist in 20, 30 years' time," Preciado tells me on a sweltering day with temperatures well over 30ºC. "We have to work for my grandchildren – not so much for us," says Preciado. I ask how many grandchildren and he beams back, "Thirteen!"

Yet sea level rise is far from the only reason to leave Gardi Sugdub.

The island is only about 350 metres long and 150 wide, so 1,450 inhabitants on this speck of land mean a population density greater than in Manhattan or London. And the Guna are living in single storey dwellings, rather than in high-rise buildings that mean everyone can get more space. One of the few buildings with two storeys is a concrete guest house where I stay, with a dock lapped by tiny waves from a flat Caribbean.

I've come here because I've read articles about the Guna plan to move to the mainland, generally written in the early 2010s when the project seemed to have momentum. There has been less coverage since of the reasons for years of delays that may help other communities improve planning.

From the air, or viewed from Google Maps, the island of Gardi Sugdub is a chaotic patchwork of grey, silver, yellow and red roofs, ringed by small quays for motorboats that zip back and forward to the mainland or ferry tourists to outer islands, spluttering out diesel fumes. Between the quays, on stilts above the water and ringing the island, are latrines. It's a stark contrast with the tourist-brochure vision of Caribbean islands further offshore, fringed by palm trees and golden sands.

Under the plan to move from the island, anyone who doesn't want to move to 'tierra firme' – the mainland – can stay behind. Guna leaders say 80 per cent to 90 per cent of those living on Gardi Sugdub favour leaving the island, especially the young. "We have to move for the sake of the children," says Ginela Salazar, a primary school teacher sitting with a group of friends as a group of children play pan pipes nearby and practise dancing on one of the small open patches of sandy ground on Gardi Sugdub. Nearby, someone is barbecuing chicken and fish, the smoke from the charcoal rising vertically into the windless sky at sunset.

The island school has 500 students, many of them from other islands in the region who live here during the week with relatives, friends or other families. After school, children crowd in the streets to play – they bang drums, draw in the sand, play volleyball, dance and twirl batons, throw marbles and jacks, or push plastic cars along the street. These toys are the only vehicles on the island. I see one albino child – the Guna have high rates of what are known as 'Children of the Moon' because of their alabaster complexions.

But none of the children are swimming – the water around here is too dirty, and they have to travel by boat elsewhere to splash around. The narrowness of the streets also means that it's hard to play games that need space. The school yard is just big enough for volleyball, but children say some of the older Guna grumble that they get struck by wayward balls that bounce off down the narrow alleys of fragile houses made from cane. Some children jog back and forward to get exercise in the yard – but it's only about ten steps before they run out of space and have to turn around. Anyone wanting to play a game that needs more space – like soccer – has to take the boat to the mainland where there is a muddy, grassy pitch.

The overcrowding is getting worse – the number of people here has roughly doubled in the past thirty years, driven by high birth rates among the Guna.

There are few trees, while just a few islanders manage to maintain tiny gardens of herbs or bananas. Pollution, with stacks of plastic littering the fringes of the island, and damage to the local corals by mining to build protective walls, all make Gardi Sugdub ever less attractive with the worsening sea level rise. Unlike the shoreline, the sandy streets are all clean – meticulously swept by hand every morning at dawn.

It is hard to keep secrets here – between the canes of many of the walls you can glimpse and hear what the neighbours are up to, meaning everyone has to whisper for privacy. Passing one house, I hear a Guna doctor softly chanting to help cure an elderly woman who has fallen sick.

The main street is a sandy path about 4 metres wide with alleys off to all sides dotted with homes with hammocks dangling from the roof beams. Clothes hang from washing lines along the street. With such crowded conditions, the coronavirus pandemic also hit the region hard in 2020 and 2021. The economy has been battered too by a halt to foreign tourists visiting the islands.

So sea level rise, like in many places on the frontlines of climate change, is just one big factor adding pressure for a move to the mainland. If sea levels continue to rise on the current accelerating trend in coming decades, it will become the dominant driver uprooting people from low-lying coastal zones and islands from the Caribbean to the Pacific and Indian Oceans.

If the UN's predictions about sea level rise are right,[2] that would make places like the Guna Yala islands uninhabitable in coming decades. Plus storms in the region may be becoming more powerful. Panama frequently suffers tropical storms even though it is south of the path of Atlantic hurricanes that wreak havoc further north in the Caribbean, Central America, Mexico and the US.

The lure of the mainland for the Guna got a boost in September 2020 when Panama's President Laurentino 'Nito' Cortizo inaugurated[3] a US$5 million electric line to power the coastal strip of villages, including the region of Nuevo Carti, the 300-house village under construction for the relocation.[4] Previously, the region has relied on electric generators or solar panels. The new electric cables do not extend offshore to the islands including Gardi Sugdub.

Videos of the event show that Cortizo invited a Guna woman, wearing bright-coloured traditional 'mola' dress including a red and yellow face mask and 'wini', bands of tiny beads wrapped on her forearms up to her elbows, to turn on the power by pushing up a lever at the end of the 44km line connecting the Caribbean coastal region over the hills to the national grid. She does so, causing a loud click and a round of applause by the president and attendees including the president's gigantic bodyguard, standing behind him in camouflage gear. Cortizo points up to a lamppost where a bulb is starting to shine beneath an overcast sky and bumps elbows with the Guna woman for a Covid-era celebration. "This is lighting lives. I have had the privilege, the honour, to witness the illumination of lives," Cortizo says in a speech.

Cortizo won office in 2019 promising to root out corruption[5] in Panama. It is an uphill task, as shown in Guna Yala in the days after Cortizo's visit. The Governor of Guna Yala, Erick Martelo, gushed in upper case on Facebook when Cortizo connected the electricity: "THANK YOU PRESIDENT NITO CORTIZO, FOR YOUR VISIT TO THE GUNA YALA REGION. WE WILL CONTINUE WORKING FOR THE DEVELOPMENT OF OUR PEOPLES!!" That work lasted less than a week, when Martelo was arrested and sacked by Cortizo after police said they caught Martelo in a car with secret compartments beneath the floor containing seventy-nine packs of drugs. Cortizo appointed Alexis Alvarado as his replacement, writing what sounded like a veiled warning on Twitter[6]: "We trust that his performance will match the expectations that the government places in his ability and honesty."

Panamanian officials say the remote Guna Yala region has long been used as a cocaine trafficking route from Colombia towards the US, with drug-laden boats threading their way through the scattered islands.

Despite the setbacks and complications of moving to the mainland, the Guna have one massive advantage for anyone moving inland – they own both the islands and a narrow strip of land stretching east to the Colombian border. That means there is no complex negotiation over land ownership that could scuttle the plan.

Apart from about 33,000 people living in the Guna Yala region, thousands more Guna live in Panama City and other parts of the nation. And, unlike many island peoples, the Guna have not been living on the islands since ancient times. In some ways, the move to the mainland will mark a return to ancestral roots. Until about 150 years ago, the Guna traditionally lived on the mainland but were steadily driven offshore by the arrival of Spanish-speakers, conflicts with other Indigenous groups, and a push to escape mosquito-borne diseases, such as malaria and yellow fever.

They picked the islands just offshore, like Gardi Sugdub, which are close to the mouths of rivers on the mainland to ensure fresh water supplies. Many of the islanders maintain farms on the mainland, for crops including coconuts, maize, rice, sugarcane and mangoes. Islands further offshore tend to have fewer inhabitants. And the islands are in many ways safer places to live than the mainland, except during storms. There are no venomous snakes and no jaguars or pumas, a few of which roam in the depths of the mainland forests.

Historically, the Guna have been highly independent, right from the start of European settlement. Apart from Spanish colonists, incursions into Guna Yala territory included a humiliating Scottish attempt in 1698 and 1699 to create a colony called New Caledonia.[7] Investments were encouraged by talk of friendly Indigenous peoples – the Guna – and the hope of setting up a shortcut overland trade route across Panama, at a time when ships had to make a perilous long journey around Cape Horn to travel between the Atlantic and the Pacific. The Panama Canal, to the west of Guna Yala, opened more than two centuries later in 1914. Many of the Scottish settlers died of tropical diseases and malnutrition. One group abandoned an early fort when they heard the Spanish were planning to attack. In total, 2,000 people died in Scotland's failed attempt to settle on the isthmus, with historic repercussions back home.

"The Darien venture was one of the most harrowing disasters to befall any nation, and the forced parliamentary union with England in 1707 was the bitter consummation of those who had dreamed of creating a Scottish empire," wrote John Prebble, author of *The Scottish Dream of Empire*.[8] In one final vestige of the failed venture, Panama in 2011 changed[9] the name of one coastal village from Puerto Escoces (Scottish Port) to Sukunya Inabaginya to reflect its Guna roots.

The Guna have long prided themselves on their fierce sense of identity and wariness of strangers in past centuries, illustrated by one grim tale related in a report by the Inter-American Development Bank[10] of 'La Bella Blanca' – the beautiful white girl.

The bank's report, examining the social and cultural impacts of moving from Gardi Sugdub, said the Guna story tells of the mysterious arrival one day long ago of the girl in a Guna village. She stayed on, the Guna cut her hair when she reached adulthood, and many of the young men fell in love with her. She married one of them, but he promptly died. Undeterred, another suitor fell for her, married her – and also succumbed. This cycle of love and death went on repeatedly despite the mounting death toll – in total, nine Guna men failed to appreciate her fatal attraction. "After that, the community decided to kill the young woman. Bad mistake," the report says. Her death in turn unleashed a deadly epidemic that killed locals in their hammocks and forced survivors to migrate westwards.

After Panama gained independence in 1903, the government tried to impose a national identity across the nation, prompting the 1925 Kuna Revolution.[11] With US backing, the Guna people managed to negotiate a semi-autonomous status within Panama, established in 1938. Despite their resilience, the move to the mainland will be a massive shift even though it is ancestral territory.

"Moving back to the mainland marks a radical change – we'll be with the forests and the rivers, not the sea," says Blas Lopez, another leader of the island community. "The goal is to avoid losing our identity," he tells me in his home on Gardi Sugdub. "But we have a big advantage because we have our own land."

Change is certainly coming for the Guna.

In 2010, Panama opened a paved road, like a roller coaster through the forest and over steep hills, linking the Caribbean port of Carti, over the water from Gardi Sugdub, with Panama City. That road, replacing a dirt track, has cut the travel time from a whole day to about two and a half hours, and has transformed the economy towards more lucrative tourism, helping to supplement incomes from subsistence fishing and farming. "Everything changed with the road, both in a positive and a negative way", Lopez said. "It has opened the region to tourism, cabins, trips to the islands. It brought drastic changes." The road also makes it easier for the Guna to seek jobs in the city, even though many who try their luck in the capital end up living in slums. Many women work sewing and selling bright-coloured traditional mola dress, as well as elaborate needlework pictures of parrots, palm trees and villages.

From Gardi Sugdub, you can see the idyllic islands with coconut palms, yellow sandy beaches and bars in the distance further offshore. On days with bright sunshine when the blue water is calm, the distant islands seem to float above the water, mirages on a reflected slice of the sky. On those islands, visitors go scuba diving on the reefs, visit 'natural swimming pools' where the water is crystal clear and about 1 metre deep and bright-coloured starfish crawl over the sand as tropical fish teem on the reefs.

It costs US$140 or more for a one-night all-inclusive stay in the islands, including transport by a jeep from Panama City – past a checkpoint where foreigners have to pay up to US$20 to enter the Guna Yala territory. The Guna are unusual in Latin America by having the autonomy to impose such checkpoints and charge fees.

Maybe moving to the mainland will mean a renewal of experiences from the distant past. Oziel Gonzalez, a Guna who works as a driver for visitors on the Panama to Guna Yala route, said that when he was a child he used to visit elderly relatives who lived in the forest. "They served us iguana," he told me, laughing, as we drove along. He said his eyes boggled when he saw the green skin on his plate. "It put me off. But if you get rid of the green skin – it tastes like chicken."

As is so often the case, it's not the big issue that preoccupies people's minds on a daily basis, more the reality of how they're going to live their lives. And yet the problems that they're facing require some bigger-scale thinking.

Elsewhere in the Caribbean, many nations focus mainly on protection against storm surges caused by hurricanes rather than on sea level rise, according to Adelle Thomas, Director of the Climate Change Adaptation and Resilience Research Centre at the University of the Bahamas and who also works for the Climate Analytics research group. "I have done so many different talks saying, 'We need to start thinking about sea level rise, we need to change'. But you do a talk and it gets written up in the newspaper and nothing actually happens," she told me. "It has to do with the difference of the short time that politicians are in office, versus the time scale needed to address these issues."

On the Guna islands off Panama, some local leaders are also trying to act by reforming harmful traditions, like the use of corals to reinforce islands, to help build resilience to storms. Eustacio Valdez Montrevo, a teacher known locally by his nickname 'Atahualpa' – the name of the last emperor of the Inca Empire in what is now Peru – says many local children don't even know that corals are tiny animals that create reefs with millions of their stony skeletons. "If we extract corals we are killing many thousands of species. They're alive," he tells me with an infectious enthusiasm sitting on a bench in his garden, one of the few on Gardi Sugdub. "If you go to a beach at night you will see lights, lots of lights. And you'll ask, 'What's that?' They're corals," he said of the tiny animals that can flicker in the dark. "Some of the children were surprised to learn about corals. They view them as rocks. But they're not. They are living things."

He tells me how he takes children on trips with masks, flippers and snorkels to visit reefs to educate them about the need to preserve corals. The hope is that they will go home and tell their parents, who are mining the corals, about the wonder of the glowing reefs.

In Guna culture, he said, the sea is viewed as a grandmother. "For many years our parents have told us that the sea is our grandmother because it is feeding us with fish," he said, arguing that was another reason to safeguard corals. By contrast, he says, the mother is associated with the land – rivers, trees and the Earth – while the heavens – the sun, the clouds, the moon – belong to the paternal side.

Interestingly, researchers have long documented how the Guna have undervalued corals. Mònica Martinez Mauri, an anthropologist at the University of Barcelona who spent months in the region in the early 2000s, found that the inhabitants of Gardi Sugdub could identify 243 types of fish swimming off their islands.[12] She said the Guna could even identify rare fish excluded from a guide of local fish compiled by the Smithsonian Tropical Research Institute (STRI), such as the flat needlefish, a fish whose sharp, thin shape inspired its name, and even a whale shark – the largest fish in the oceans that was spotted near an outer island in the 1990s. In total, she found that the locals could identify 80 per cent of the species of fish listed in the STRI guide for the region, as well as many rare visitors. Yet despite their encyclopaedic knowledge of fish, the Guna could only identify 1 per cent of the more than 100 types of corals when shown samples or photographs. Worse, she wrote, "Most of the inhabitants of Gardi do not recognize the different species of corals and refer to them using the generic term 'akkua' – literally meaning 'rock'." Their belief that the foundations of the islands are simple rock rather than living organisms threatens the Guna islands and fish stocks. "Biologists believe that the domestic waste thrown into the sea and the extraction of corals to reinforce the islands are the main causes of the degradation of coral reefs in the region," she wrote.

The damaging practice of mining corals and filling up islands probably began in the early twentieth century; it was also used on the mainland to help build an airstrip on formerly swampy land by the port of Carti. Outer reefs, vital for seafood ranging from fish to lobsters, also act to break the power of waves sweeping towards the coast.

Around the world, global warming is harming corals from the Caribbean to the Great Barrier Reef off Australia. When temperatures rise, the corals can go a ghostly white, expelling the bright algae which live with them. These bleaching events, which can kill corals, are becoming ever more frequent as ocean temperatures rise. The IPCC has repeatedly sounded the alarm about the fragility of corals as the world's average surface temperatures rise. Global temperatures are already about 1.2°C above pre-industrial times. "Coral reefs would decline by 70–90% with global warming of 1.5°C, whereas virtually all (more than 99%) would be lost with 2°C," it stated in a 2018 report.[13]

At the same time, industrial carbon dioxide building up in the atmosphere is tending to make the oceans more acidic, in what some scientists call a 'silent storm', making it harder for corals and creatures such as crabs or shellfish to build their skeletons. That in turn could reduce the amount of material available to build islands. In healthier oceans, corals and mangroves can help islands grow to keep pace with small amounts of sea level rise. All these impacts of warming, resulting from emissions abroad, are compounding the problems caused by the Guna practice of coral mining.

Hector Guzman of the STRI has argued that the damaging Guna practice of mining corals had often been overlooked or "treated with indulgence"[14] by outside researchers. Living coral cover in the Guna Yala region declined 79 per cent in the last thirty years of the twentieth century, he wrote. He and his team measured 20km of sea walls built with mined reef corals – equivalent to 16,000 cubic metres – and recorded an increase in island surface area of 6.23 hectares in the region linked to the coral filling. In other places, unprotected coasts eroded.

"Coastal erosion has increased as a result of the lack of a protective natural barrier and ... a local increase in sea level. Coral-mining and land-filling practices to accommodate population expansion and mismanagement of resources have significantly modified the reef ecosystem and will have serious long-term consequences," his team wrote in a study published in the *Journal of the Society for Conservation Biology* in 2003.

The Inter-American Development Bank says that families on Gardi Sugdub have steadily expanded the coastal shorelines with corals. "The island observed today may be, in area, the double of the original," according to a 2018 report.[15] "In line with oral traditions, on arrival in the islands the Guna found mangroves, which were cut down to use the wood as fill. After the collapse of mangroves, corals have been used," the report detailed.

Natural corals are also a barrier against storms and even tsunamis in the region. In 1882, a tsunami triggered by a major earthquake killed perhaps 100 people on the islands.

Carlos Arenas, author of a report[16] for Displacement Solutions, a non-governmental organisation which helps the Guna, said that coral mining had "exacerbated their vulnerability, as the coral reef acts as a natural barrier against storms and sea tides". But he said that the mining was a response to forces of climate change beyond their control, caused by nations emitting greenhouse gases. "As negative as this practice [of coral mining] might be, it doesn't make the Guna responsible for climate change, or for the rise of sea levels, as some people have suggested," he wrote. Displacement Solutions has long faulted the Panamanian authorities for failing to ensure that the Guna can move quickly to the mainland. "There is a serious risk that when the sea level reaches a trigger point, or when a natural disaster strikes, they would probably be forced to move to the slums of Panama City where they would lose their livelihoods and traditional way of life," it wrote. "This would clearly be a great loss not only for the affected communities themselves, but also more broadly to the Indigenous culture of Panama."

Panama insists that it is now taking urgent action. It issued an updated plan for fighting climate change to the UN in December 2020 that seeks to promote a green economic recovery after the pandemic. And it includes a spotlight on the plight of Guna Yala. "In terms of sea level rise there are already critical situations among populations of Indigenous peoples. A critical case is the situation in the Guna Yala region," it says.[17]

Ligia Castro de Doens, Director of Climate Change at Panama's Environment Ministry, said the government was working hard to help the Guna people. "Two of the islands have already sunk below the sea," she told me in 2021. "They are losing part of their territory." She also noted that the government was building new homes for those relocated. She deflected criticisms that the process was too slow, commenting that it required time to build consensus on all sides to push through an emotionally wrenching project to move. "There is also resistance by some to move to the mainland – they have lived on the islands for generations," she said.

As part of wider efforts to combat climate change, she explained that the government was also working to restore coastal wetlands including mangroves, all along the nation's coasts. "Wetlands are the first line of defence against coastal erosion," she said. "There will be more green jobs," she said, pointing to plans to expand coconut groves along the northern coasts that can provide both food and roofing materials for traditional homes.

She also reflected that the problems of Guna Yala, with its closeness to the sea, were mirrored in many parts of Panama. "Panama is a very narrow country. We can drive from the Atlantic to the Pacific in an hour. In practice our climate is that of an island. The winds of both the Pacific and the Atlantic buffet us constantly," she said. In response, Panama is stepping up education about the value of protecting the environment as part of the sustainable development goals. "A programme of environmental education is being carried out in all communities, both the Indigenous and non-Indigenous ones so that they know the vulnerabilities of the country and can involve nature in the solutions. There is more vigilance, for forests, and also to ensure that corals will not be affected," she explained. She said that Panama reckoned that it was, overall, the third "carbon negative" country in the world alongside Bhutan and Suriname, meaning that its forests and other natural resources absorb more greenhouse gases than those emitted by human sources such as cars, factories and power plants.

The national climate change plan submitted in December 2020, part of the Paris Agreement adopted in 2015, says Panama will step up actions to limit its use of fossil fuels in energy and to preserve forests in the coming decade.

In contrast to corals, the Guna have been more successful in safeguarding forests – on one stretch of the route from Panama the Guna Yala territory is an impenetrable forest on the left-hand side of the road, while the right-hand side is under local Panamanian control and has been planted with commercial pine trees or cleared for cattle ranching.

"We can be guardians of the forest. We are helping to preserve the lungs of the world. We are bringing carbon to the world," Guna leader Lopez said.

The power of nature and the encroaching forests is visible on the incomplete buildings vital for the move to the mainland. The long-promised hospital is mired in legal disputes and not on the government agenda, according to Dilion Navarro, a local Guna leader. Best progress in 2021 is being made on the 300 houses on a 17-hectare plot first cleared by the Guna, on land donated by local families. "The move is for the start of 2023, with the construction project completed at the end of 2022," Navarro told me.

The Panamanian Ministry of Housing revived the house-building project in October 2020 after coronavirus forced a break. At the entrance of the site are large signs making face masks obligatory, demanding regular hand washing and social distancing. "The residential area which is being built for more than US$9.3 million has the aim of moving families from Gardi Sugdub," it states. Some streets have been laid out, with concrete foundations and drainage pipes laid.

Of the total 300 houses, 295 houses will be 41 square metres in area, while five will be for people with disabilities and will be slightly larger, 43 square metres, it said. "The Guna will benefit from two neighbourhood parks and one sports ground, a plant for water treatment, streets, pavements, services of electric light and drinking water." And the new village, Nuevo Carti, will also have halls where Guna traditionally meet in the evenings, built with traditional wooden materials. Once completed, families would benefit from "greater security and quality of life", the Panamanian Ministry of Housing said.

Gardi Sugdub resident Magdalena Martinez Allen expressed worries when we met that the identikit homes will be too small for the large families of the Guna people. In Guna tradition, a married couple moves in with the bride's family. Times are, however, changing. Many younger people want a place of their own, away from their parents or in-laws. "We are pioneers. This island is the first to make this work of moving to the mainland," she says. "We need more space."

Like many Guna, she says the community needs help. "You've robbed a lot of land from us," she tells me of the legacy of European settlers in the Americas. "All of it was ours from the north of America to Patagonia. We lived happily. You should at least support us. That's what we want."

The Inter-American Development Bank, which has supported the construction[18] of the US$10.8 million school and other aspects of the move to the mainland, said that those wanting to live in the new village were mainly from Gardi Sugdub, with some from the diaspora living in Panama City. It cites the potential risks as including "possible conflicts with other communities or authorities in Guna Yala, who may resent the fact that Gardi Sugdub is the community receiving most resources for moving".

And there are many other communities at risk in Guna Yala who may want to follow the move to the mainland by Gardi Sugdub, if it succeeds. People on a string of other Guna islands such as Playon Chico, Nargana and Corazon de Jesus are all using corals to protect from floods, the Panamanian government says in its 2050 plan for combatting climate change.[19] "The sun is getting closer every day," it quotes the inhabitants of Gardi Sugdub as saying of hotter conditions.

On Playon Chico, which has a population of about 3,000 people, it says there is a lack of vegetation and floods have eroded homes. On the island of Nargana, it says community leaders have discussed the idea of moving. "They have a preliminary site ... as a possible option," it says. Similarly, on the island of Corazon de Jesus, the community "owns land and property on the mainland to move".

But such projects will inevitably be costly, and Panama lacks resources.

Of the projects on the mainland opposite Gardi Sugdub, the school has long been the most advanced. Diomedes Fabrega, the deputy director of the school on the island of Gardi Sugdub, told me when I visited that 95 per cent of the new school on the mainland was finished, the air conditioners are there, they just lack electricity, and toilets and wash basins have also been plumbed in – even though there is no water.

When I visit, a bat flies out of a refuge in the roof of the big indoor gymnasium, a lizard skitters across the floor of the school, and spiders lurk everywhere. One option is to open the school before the houses are completed and transport the 500 children to the mainland and back to the island daily, Fabrega said. But it would be costly, requiring a fleet of five extra boats, he reckoned, as well as buses from the port to the school. Each of the new boats would cost about US$25,000.

And many Guna are rightfully wary of travelling out at sea when thunder clouds loom, for fear of lightning strikes. I delayed a trip to the mainland from Gardi Sugdub when the weather darkened.

However, at least the school marks some progress, which is more than can be said for the hospital – begun in 2011 with a budget of US$11 million under a former government, construction was abandoned amid legal disputes in 2016. Big green stains are creeping like fingers up the walls of the hospital and orange rust is spreading everywhere. "A monument to oblivion," Panamanian TV station Telemetro[20] dubbed it in a report in late 2020, showing the forest slowly engulfing the building.

"It's been a total waste," Alejandro Barranco, the administrator of the port of Carti a few hundred metres down the road, told me as he showed me round the overgrown site. "Millions of dollars have been squandered. If this was Panama City it would have been finished by now."

The hospital is part of a labyrinthine legal dispute[21] between the government and building company Omega Engineering, still rumbling on in 2021 and limiting healthcare for the Guna during a pandemic. Both Omega and the Panamanian government accuse the other of corruption and violating contracts. The legal tangles and multiple setbacks mean the Guna are a lesson for other nations – about the cultural, social and economic complexities of moving as seas rise. The decade-long plan to move to the mainland is demanding endless patience.

Preciado, the village elder on Gardi Sugdub who saw sardines in his house, says that he feels like the captain of a sinking ship whose job is to ensure that the passengers and crew leave safely, no matter how long it takes. "The captain is the last to leave," he said.

7

SEAL ROCKS AND ANCIENT OAKS: THE MYSTERY OF THE FALLING BALTIC SEA

I have seen myself what was once firm land, become the sea: I have seen earth made from the waters: and seashells lie far away from the ocean, and an ancient anchor has been found on a mountaintop.

Ovid, *Metamorphoses XV*

In Sweden the level of the Baltic Sea is falling relative to the land – starkly at odds with global sea level rise. Scientists now know that the region is rebounding after the end of the Ice Age 10,000 years ago lifted a vast weight of ice off the land, but it took them centuries of detective work to figure out why land constantly emerged from the sea and why ports dating back to Viking times grew ever shallower. Anders Celsius, the inventor of the temperature scale that bears his name, played a key role in starting to solve the mystery in the eighteenth century, with novel data mining based on 'seal rocks'. In the nineteenth century, British geologist Charles Lyell used an ancient oak tree near Stockholm as a botanical clock to understand the rising land.

HUNTING FOR A GREY BOULDER that is a landmark in the history of changing sea levels, I park my car at the end of a bumpy track near the Baltic Sea and look for a path down to the shore through a Swedish pine forest. There's no one around to ask the way on this remote Iggön peninsula about 175km north of Stockholm, just a few seagulls flapping overhead, and no sign of a way through the trees. But it's a pleasant warm July day with a scent of forest sap in the air combined with salt from the Baltic Sea. I've come to find an almost forgotten rock near the shoreline, guided by precise geographical coordinates sent to my phone – latitude 60°52'56" north, longitude 17°15'47" east – by Swedish scientist Martin Ekman, an expert in sea level changes and associate professor of geophysics at Uppsala University.

While many of the world's coasts are under threat as seas rise, here the sea is bizarrely falling relative to the land, in some places by a centimetre a year. That means a metre every century, a staggering 10 metres in the millennium since Viking times.

Scientists now know that the land here, and in some formerly ice-smothered parts of the globe such as Alaska and northern Canada, is rebounding after the ending of the last Ice Age 10,000 years ago lifted the enormous weight of an ice sheet more than 1km thick off the land. The region is still rising, like a foam mattress that takes a while to reshape after you get up off it. And the Earth's mantle deep below is a gooey mass that takes millennia to reform after it was weighed down by ice. Yet well into the nineteenth century no one knew about Ice Ages and were baffled about what was happening: was the land rising, or the sea falling, and why?

The rock I'm trying to find is Exhibit A in a centuries-long detective story about the rising land and how Sweden's Anders Celsius, a professor at Uppsala University and an astronomer and physicist, was the first in the eighteenth century to start measuring the rate of change along the coast.

It was a pioneering, vital step in figuring out what was going on, and marked the start of more than a century of work by the some of the smartest minds in Europe to solve the puzzle of the rising land and the falling sea.

Best known for inventing the temperature scale that bears his name, Celsius rarely gets the credit he deserves for his groundbreaking work on sea levels, based on mining data from the natural world. One of his greatest achievements was identifying the rock I'm looking for.

I stare at my phone: Google Maps predicts a roughly 1km slog through the forest to the boulder, but can't tell me how to get there. I set off, fending off cobwebs, mosquitoes and flies, with branches whipping needles as I push past.

In the 1560s, almost 200 years before Celsius came on the scene – the decade when Galileo and Shakespeare were born and Sweden was on its way to becoming a regional superpower in rivalry with Russia – Iggön was an island, rather than the 4km peninsula it is now. Records from the time show that the rock I'm looking for was owned by a farmer, Rik-Nils (Rich Nils) and it was clearly just above the water line – local seals could clamber on top to sunbathe in the brief Nordic summers. Crucially, Celsius found this rock was listed on historical documents from the 1560s as a 'seal rock'. When close to land, such 'seal rocks' were often marked on documents of land sales because they were valuable – hunters could harpoon the mammals for food. By contrast, random rocks in a forest, or those jutting up too high from the sea, were worthless and not recorded. So, in the middle of the sixteenth century this rock was a flattish-topped boulder where seals could slither out of the sea with their flippers to bask. If the rock jutted up too much, seals would be unable to get on top. Ekman, who texted me the coordinates, has previously told me that Celsius' seal rock is now high and dry in the forest. That seems extraordinary and, if true, it makes it a 500-year-old rock clock, indicating the changing sea levels.

I walk on over moss and try to avoid tripping on tree roots.

Throughout Swedish history, the rising land along the Baltic coast was a puzzle – it has left Viking ports high and dry, drained sea moats that once protected castles and forced relocations of entire medieval towns towards the sea. One tide level gauge carved into a rock on the Åland Islands in the middle of the Baltic Sea, an autonomous region of Finland, is now incongruously marooned on land; it was designed to guide ships into port in the nineteenth century but is now hard to see behind rocks, reeds and wildflowers, near a modern bridge.

And the rising land still has consequences. It creates new land – by some estimates about 7 sq km every year for Finland alone – while northern ports such as Luleå in Sweden need to keep dredging to ensure ships can dock. Now, the global rise in sea levels is partly offsetting the local fall, gnawing away at a Nordic geological advantage over most of the world. Celsius' work inspired other scientists to visit Sweden in the nineteenth century, including Charles Lyell, a British geologist. Lyell's work uncovering the apparently infinitesimal processes of geological change over millions of years helped his friend, Charles Darwin, to develop his theory of evolution. It also challenged the common belief among Christians that the Earth was only a few thousand years old.

For most people living around the Baltic Sea, falling sea levels were just a baffling fact of life. Some Christians speculated that the level of the Baltic Sea was falling in a lingering sign of retreating waters described by the Biblical story of a flood sent by God to punish wayward humanity. Devout Christians reasoned that the water might be still gurgling away into the ground after Noah built an Ark to save his family and creatures – from aardvarks to zebras – two by two. But that theory had obvious flaws: if the water was receding in the Baltic Sea through some enormous hidden subsea plughole, why wasn't it also falling everywhere in the world's interconnected seas? And if there was a difference in sea levels between the Baltic and the North Sea, you would expect strong currents through the narrow straits between Denmark and Sweden, past what is now the Øresund Bridge linking Sweden and Denmark, and immortalised in Scandinavian noir crime series *The Bridge*. But there are no abnormal currents. Other theories were that vast springtime thaws from rivers flowing into the narrow Baltic Sea might be carrying silt and rocks, somehow building up the shorelines. The Baltic is ringed by Sweden to the west and north, and then Finland, Russia, the Baltic States, Poland, Germany and Denmark in the east and south.

Even now, the rising land is still a source of wonder, and uncertainty, for many Swedes. This is the home of Thunberg, who has inspired a global movement by accusing governments of inaction on climate change.

"It can still be confusing," Anna Rutgersson, a professor of meteorology at Uppsala University's Department of Earth Sciences, told me. "There is a lot of talk about climate change and melting ice and sea level rise. But due to the land uplift, there is no increasing sea level that people can experience."

With an unexpected thrill, I eventually spot the elephant-size boulder between the pines by the shoreline as I reach the coordinates sent by Ekman, who identified the rock in 2012 after it had been largely forgotten since Celsius' time. "You're probably the first person to visit since me," he'd told me. He described the rock in a report[1] for the Summer Institute for Historical Geophysics, and he is a one-man mine of information about the Baltic coast.

The almost submerged boulder of 500 years ago is now high and dry in the forest, mottled by moss with a sprinkling of leaves and pine needles on top. Weirdly for a former 'seal rock' it is about 10 metres back from the shore. A few sparrows chirping on top of it flit away into the trees as I approach. At a rough guess, I reckon it has probably risen about 4 metres in the 500 years since Rich Nils and his family used to come here to hunt. It's no longer a seal rock, but a perch for birds and butterflies. And as Ekman warned me, it's pretty anonymous, even for a rock. "From the historical point of view, it's a very interesting rock. But there is nothing to see – it's just a rock in the forest, no inscription or engraving. There are no other marks on it," he said. He reckons it is "probably identical with Rich Nils' seal rock used by Celsius". There are no others in this area matching Celsius' descriptions – but of course it's hard to be sure.

Roll back almost 300 years to the 1730s, when Celsius decided to use science to understand what was going on along the coast. Like many people of the time, his hunch was that the sea level was falling – it seemed inconceivable that the land itself could rise. But he had a problem that persists in modern climate change science – how do you measure shifts in sea levels or temperatures, or the extent of glaciers or any other such changes when there are no historical benchmarks to guide you? Understanding ice in Antarctica, for instance, is a detective story in progress – the continent was only first spotted by sailors in 1820. Nevertheless, measuring change is vital, not least to persuade people who don't believe it's happening. This century, public acceptance of man-made climate change has widened, for instance, as temperatures have risen compared to historical records.

For the Baltic Sea region, most written records were not much use beyond generalities. In one of the earliest about land uplift, Ekman quotes[2] Swedish historian Elias Brenner writing in 1694:

> From old times one has noticed and experienced how large bays of the sea gradually turn into land. Where people in several places 70 or 80 years ago could sail freely with small boats, now every year hundreds of loads of hay are gathered. ... One has also found that the waters up to one Swedish mile (about 10km) or more off the coast gradually become more unsafe. This has been experienced by several navigators whose ships have run aground and been damaged, where never before any cliffs or grounds have been detected.

So Celsius, whose face is best known from a dark portrait showing him with an enigmatic smile wearing a curly grey wig down to his shoulders, struck upon the idea of poring through historical records to find some historical benchmarks for the coast.

A polymath from a leading academic family who was appointed professor of astronomy at Uppsala University at the age of 29, Celsius' eureka moment came when he discovered the records of seal rocks. Celsius found four seal rocks in historical records, two in Sweden and two in Finland, including the one at Iggön owned by Rich Nils. The records show Rich Nils set a fire on the rock in the 1560s, apparently to help crack off bits to make the top smoother and more attractive to seals. His sons formally bought the island from the crown and received a taxation certificate from King Johan III in 1583. The rocks were declared worthless on later taxation documents because they were too high out of the water for seals.

Celsius sent a local mathematics teacher, known as Rudman, through this same area to measure and draw the rock on Iggön in 1731. His drawing from that year showed that the water was lapping around the base of the rock. Based on Rudman's measurements, Celsius estimated that the sea had fallen 237cm from Rich Nils' time in 1563 to 1731, an average rate of 1.4cm a year.

That was a scientific breakthrough – the first estimate of the rate of sea level change based on the eighteenth-century equivalent of 'data mining' – turning raw data into information. It's in the right ballpark, though above modern estimates of about 0.8cm a year in the region. But before that no one had been able to give an estimate of the rate based on anything but guesswork. It needed the skills of Celsius, whose epitaph in a church in Uppsala reads: "Clear sense, honest will, careful work and useful learning".

Lacking knowledge of ice ages or of rates of sea level change elsewhere in the Baltic, Celsius concluded that the water level of the Baltic Sea was receding, rather than the land rising. That turned out to be wrong, oddly an echo of the little-known upside-down approach to his original temperature scale. On that first scale, published in 1742, Celsius set freezing point at 100 degrees and boiling point at zero. He reasoned that it would be confusing to jump to negative numbers for freezing temperatures that were frequent in winter in Sweden, simpler to call a cold day '110 degrees', for instance, rather than 'minus 10'. But scientific opinion was against him – the scale was reversed to the numbers we now know.

After Celsius' time, the rock was largely forgotten until Ekman came across a sketch of Iggön island, with the seal rock marked, among Celsius' notes left on his death in 1744. Ekman also wrote a biography of Celsius[3] after finding that many people knew very little about the man behind the temperature scale – he calls him "the unknown man with the well-known name". Ekman says the rock in the drawing made by Celsius' colleague Rudman doesn't exactly match the one in the forest. Still, Rudman might have drawn the rock from memory back in his office, rather than while sitting here in the musty forest. On Iggön I walk around for an hour or so looking for other candidate boulders but also can't find any that come closer to matching Celsius' description. The receding seas are still a feature of Iggön, which is a leafy peninsula with forests, farms and summer cabins. One resident showed where her grandfather, as a child, used to leave his clothes on a flat rock on a former shoreline when he changed into his swimsuit to go for a dip in the sea. The water's now 100 metres away. And on Iggön, Rich Nils lives on in folk memory. Several locals told me that every time someone digs a ditch or a foundation for a building, they dream – so far in vain – of turning up some buried treasure from the sixteenth century left by Rich Nils.

To get a stark impression of the scale of change in Sweden since Viking times, travel to the railway station at Upplands Väsby, a commuter town near Stockholm. The area around the station is a flat, unexciting place, with a parade of buses swooping in and out to deliver or pick up commuters, alongside coffee shops, a modern library and fast-food restaurants. But go down the station steps to cross below the tracks and you enter a magical tunnel. Far from being underground, it makes you think you are under water – there is a huge watery blue mural imagining how this spot was below sea level 1,000 years ago. Fish swim alongside you and, as you look up, you can see the hulls of wooden Viking ships painted with oars dipping into the water. On one you can see shields hanging over the side and spears standing tall in the ship. A string of bubbles pops up from the seabed. One man seems to have fallen overboard and is drifting with the current, apparently drowned. It's an arresting work of perspective to illustrate land uplift at work.

"The land here was probably 5 metres lower in Viking times," said Tove Stjärna, a local archaeologist. A 2019 report she co-authored shows that most of what is now Upplands Väsby's town centre would have been under water 3,000 years ago, and water channels would still have snaked through the town centre in Viking times. At the bottom of the stairs in the station, an elderly man stops by the mural and reads a sign, put up by the local authority. "Did you know that this was the seafloor in Viking times?" it asks. "Viking ships sailed right here," it says. In stark contrast now, about 100 trains a day pass overhead.

Even more striking, a few hundred metres north of the station, away from the town centre, is Runby, a wooded area where Vikings once lived by the former shoreline. On a stone in the wood, a woman called Ingrid had a looping runic inscription carved around the year 1050 as a memorial to her husband and two sons. They've evidently died, but she gives no hint of their fate. Her inscription may also be the earliest written evidence of sea level change in Sweden. "Ingrid had the loading bridge made and the stone cut after Ingemar, her husband, and after Dan and after Banke, her sons," it says, according to a translation on a sign alongside. The runes are etched into the stone and highlighted in red paint. "They lived in Runby and owned a farm there. Christ help their souls. This will be in memory of the men, for as long as people live." The sign beside the rock says that the word 'loading bridge' indicates it was a sort of wharf for getting cargo onto and off ships 1,000 years ago, allowing ships to make 'continued sea travel' through bays along the coast that might have been navigable as far as Uppsala.

After Celsius made the first estimate of the rate of change, it took much more work to reach the modern understanding that the land is rising around the Baltic Sea, rather than the sea level falling as Celsius believed. Part of the credit for that discovery also goes posthumously to Celsius, because he laid the foundations for better understanding by marking rocks at sea level. Such benchmarks helped to measure how quickly rocks rose out of the water in subsequent years, and showed rates varied along the coast. The marks were also a better gauge than informed guesswork based on seals' sunbathing preferences in the sixteenth century.

Celsius had a horizontal mark about 30cm long scratched on a rock with the date – 1731 – etched above the line on Lövgrund Island, south of Iggön. It's now known at the Celsius Rock and has marks of subsequent years, including 1831 and 1931, inscribed ever lower by generations of scientists. "There are years scrawled all over it now," said Sture Sundin, a local boat owner who took me out one day from the mainland. "It's amazing how far the water's fallen in just a few lifetimes."

Sundin and I weren't as scientific as Celsius – our main bit of equipment was a 2-metre folding household measuring stick that he found in his boathouse. After mooring his boat, Sundin gamely clambered up the rock, unfolded the measuring stick to its full and lent precariously out over the water as I put up the camera tripod in the chill water, luckily without slipping on the seaweed. We found the water is now about 2 metres below the 1731 mark – an average fall of about 0.7cm a year in almost two centuries. That's pretty close to scientists' estimates, even though we didn't take account of the tiny local tides, winds, or the fact that the rock face isn't quite vertical. Each of those factors could skew our garden-shed science.

Some experts were uneasy about Celsius' conclusion that the sea level was falling. John Playfair, a Scottish mathematician and geologist who was a professor at Edinburgh University, argued at the start of the nineteenth century that the land was rising, rather than the sea falling. He reasoned that a fall in water levels in Sweden would mean displacing the entire ocean since the waters are interconnected – the Baltic to the North Sea, the North Sea to the Atlantic, and so on around the world. But he acknowledged it was hard to believe that the solid ground might move, rather than liquid water.

"The imagination naturally feels less difficulty in conceiving, that an unstable fluid like the sea, which changes its level twice every day, has undergone a permanent depression in its surface, than that the land, the terra firma itself, has admitted of an equal elevation," he wrote in his 1802 work *Illustrations of the Huttonian Theory of the Earth*.[4] "In all this, however, we are guided much more by fancy than reason; for, in order to depress or elevate the absolute level of the sea, by a given quantity, in any one place, we must depress or elevate it by the same quantity over the whole surface of the earth, whereas no such necessity exists with respect to the elevation or depression of the land." He added: "To make the sea subside 30 feet all round the coast of Great Britain, it is necessary to displace a body of water 30 feet deep over the whole surface of the ocean. The quantity of matter to be moved in that way is incomparably greater than if the land itself were to be elevated." And he noted that there were many signs of ancient shifts in the Earth that were hard to explain. Anyone digging beneath the Botanical Gardens in Edinburgh, which he reckoned was more than 40 feet above sea level, would find sand and the odd ancient shell, reminiscent of a beach.

All of this was borderline heresy for many Christians in an age when many accepted biblical teachings and thought the Earth had been created 6,000 years ago. How could a beach be stranded high above sea level in such a short time? Wasn't it more likely that a cataclysm, like the aftermath of Noah's flood, had left shells high and dry? Playfair said geologists should have the same freedom to come up with ideas about the planet as Nicolaus Copernicus, the sixteenth-century Polish astronomer who displaced the Earth from the centre of the universe by finding that the planet orbits the sun. Otherwise, Playfair contested, everyone would still wrongly be convinced that the Earth was flat and the centre of the universe.

The next step in the Baltic puzzle – finding that the rate of land uplift differed from place to place – was solved with the help of an oak tree near Stockholm beside a lake connected to the Baltic Sea. The oak is still standing in a park on the outskirts of the city. Like Celsius' seal rocks, understanding the oak required an insightful bit of data mining from nature, this time by British geologist Charles Lyell in 1834 when he was on a trip to Sweden trying to solve the mystery himself. Lyell's lecture[5] in 1834 contains enough information, with maps and diagrams, to go and find the oak even though the land has since risen.

Lyell struck on the idea of studying ancient trees, starting from his observations that oak trees don't grow in the sea and acorns don't germinate in salt water. So, knowing just two things – the age of a tree near the shoreline, and the height of its base above the water – could give an estimate of the maximum rate of land uplift. If the land is rising fast, there would only be young trees near the shoreline, with older ones further inland, higher up the slope. On the other hand, if the land is rising very slowly, then older trees could be close to the shore. In a park near Stockholm, Lyell wrote that he had studied an old oak tree with a base about 8ft (243cm) above the shore and that a Mr Stroem, the Swedish Keeper of the Royal Woods and Forests, told him that the oak "cannot be less than four centuries" old.

Those numbers implied that the land could not be rising more than about 2ft (60cm) a century, or 6mm a year. Lyell was a bit more cautious in his formal back-of-the-envelope botanical conclusion in his lecture in London, saying that the rise of the land was "very slight", perhaps 10in (25cm) a century. "It is improbable from what is known of the habits of the oak in this country, that the ... oak grew close to the water's edge originally; and if its base be now only eight feet above the mean level of the lake, it is clear that the rise in each century must have been very slight, although it may undoubtedly have amounted to ten inches in a hundred years, which would accord with the estimate of the best-informed scientific men in Sweden, in regard to the gradual rate of the rise of land at Stockholm," he said.[6]

My visit to the oak in 2019 takes place on a blazing summer's day, when it's 30ºC in the centre of Stockholm. It's a Friday so outside the Swedish parliament, where teenage activist Thunberg began her

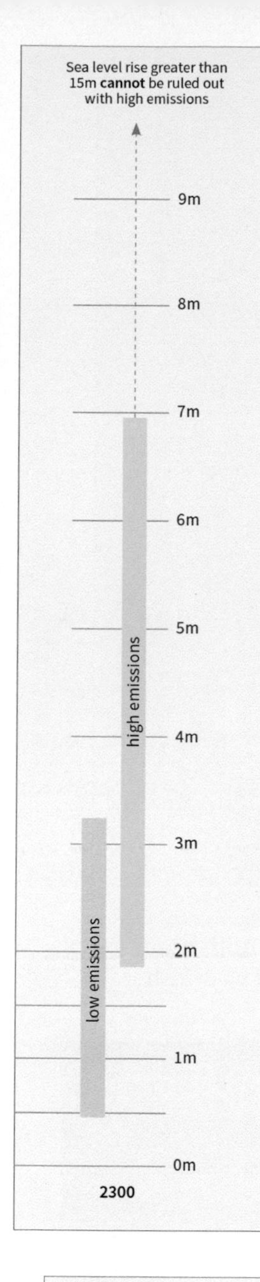

Left: Sea levels could rise by up to about 7 metres in 2300 (pink) with high greenhouse gas emissions, but deep cuts in emissions could limit the rises to less than a metre in the best case (blue). But if ice sheets on Antarctica and Greenland start to disintegrate "sea level rise greater than 15m cannot be ruled out with high emissions". (Based on a graph by the IPCC, August 2021)

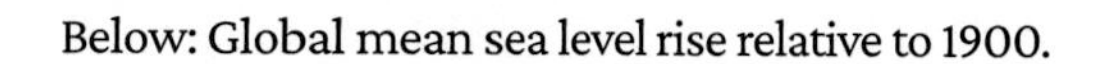
Below: Global mean sea level rise relative to 1900.

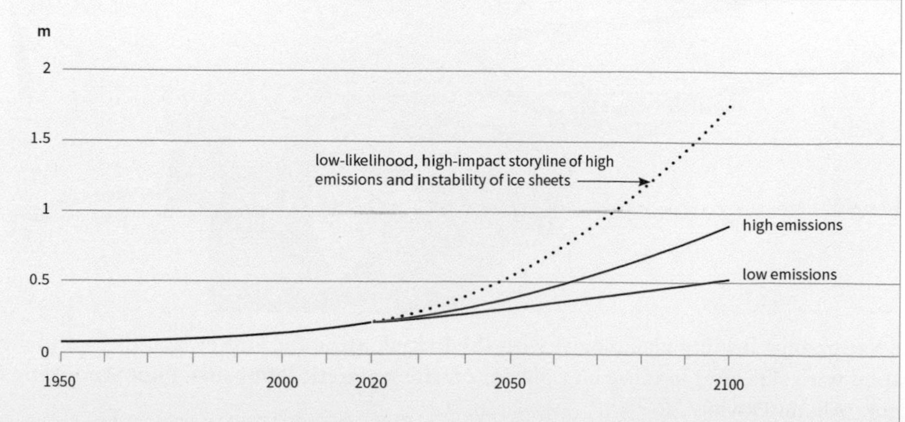

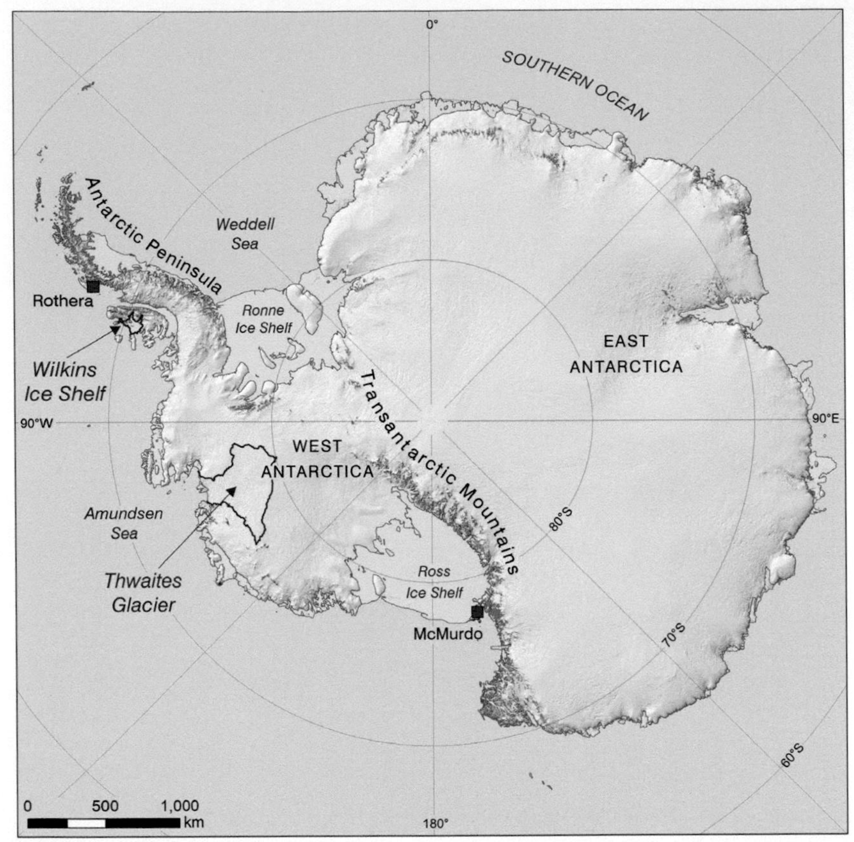

Antarctica, showing the Wilkins Ice Shelf, the Thwaites Glacier, British Antarctic Survey's Rothera base and US McMurdo station. (Map produced by British Antarctic Survey)

David Vaughan, a leading glaciologist with the British Antarctic Survey, stands by a plane equipped with skis after landing on a glacier on the Antarctic Peninsula. Pilot Steve King on the roof. (Alister Doyle)

The vast Thwaites glacier in Antarctica meets the Southern Ocean. Thwaites is the focus of scientific research into the quickening thaw of the West Antarctic Ice Shelf that could push up sea levels. (David Vaughan, British Antarctic Survey)

A view of the shattered parts of the Wilkins Ice Shelf, off the Antarctic Peninsula. (Alister Doyle)

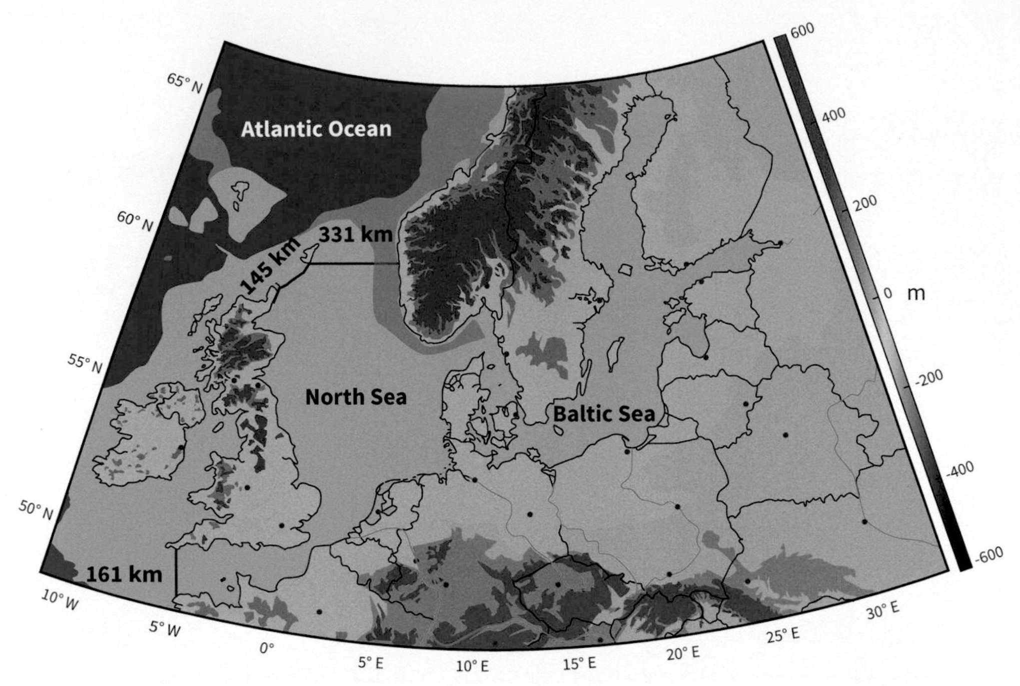

A map of Europe showing proposed mammoth dams between England and France and Scotland and Norway that would seal off the North Sea and Baltic Sea from the Atlantic to isolate them from global sea level rise. (Redrawn by Geethik Technologies, India, based on a map produced by Sjoerd Groeskamp)

The Dutch Sand Motor, a vast pile of sand and gravel dredged up from the seabed and placed on the beach at Ter Heijde in 2011. The sand is gradually washing along the coast, helping to reinforce the coastline, build dunes and limit risks of floods. (Aerial picture by Rijkswaterstaat)

Simione Botu, Chief of Vunidogoloa village in Fiji, looks out over the sea from a beach where his childhood home once stood. It was washed away by coastal floods, forcing him to move inland, twice. (Alister Doyle)

Botu stands by wooden stumps sticking out of a beach in eastern Fiji that were once the foundations of his childhood home: it was washed away by storm surges, forcing him to move inland. (Alister Doyle)

Botu built this house after his childhood home was washed away, forcing him to build about 50 metres inland. But worsening sea floods rotted the foundations of his second house, and he moved to a third home on higher ground. This third home was built with help from the Fijian government in 2014 on a hillside about 1.5km from the sea. His village of Vunidogoloa, home to about 150 people, was the first relocated in Fiji because of climate change and worsening coastal floods. (Alister Doyle)

A wire cage containing rocks lies along the coast near the Fijian town of Savusavu as part of new protections from floods caused by cyclones and rising sea levels. (Alister Doyle)

A view of New Vunidogoloa village in Fiji. The village is the first in Fiji relocated inland from the shoreline to escape the worsening impact of floods and rising sea levels. (Alister Doyle)

Children in New Vunidogoloa village in Fiji. (Alister Doyle)

Iceland held a ceremony in 2019 to mourn the demise of the Ok glacier, Okjökull, near Reykjavik. These images show the glacier atop Ok in 1986 (left side, lower image) and in 2019 (left side, top image) when it was judged to have melted away. (NASA Earth Observatory images by Joshua Stevens, using Landsat data from the US Geological Survey)

Right: A plaque placed on a rock at the top of Ok volcano near Reykjavik, Iceland, to mark the disappearance of the glacier that had previously capped the 1,198-metre peak. (Rice University, Texas)

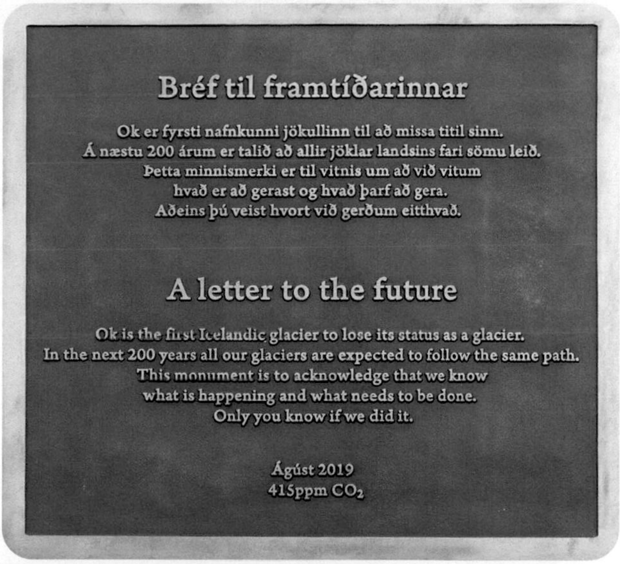

Bréf til framtíðarinnar

Ok er fyrsti nafnkunni jökullinn til að missa titil sinn.
Á næstu 200 árum er talið að allir jöklar landsins fari sömu leið.
Þetta minnismerki er til vitnis um að við vitum
hvað er að gerast og hvað þarf að gera.
Aðeins þú veist hvort við gerðum eitthvað.

A letter to the future

Ok is the first Icelandic glacier to lose its status as a glacier.
In the next 200 years all our glaciers are expected to follow the same path.
This monument is to acknowledge that we know
what is happening and what needs to be done.
Only you know if we did it.

Ágúst 2019
415ppm CO_2

Below: Svanhvít Jóhannsdóttir, a guide and park ranger, on the Virkisjökull glacier in southern Iceland. (Thorsteinn Roy Jóhannsson)

Satellite image of Tarawa atoll, Kiribati, one of the world's lowest-lying nations in the Pacific Ocean. Tarawa atoll is about 35km long and is home to Ioane Teitiota, who lost a landmark bid to become a 'climate refugee' in New Zealand. (European Space Agency, contains modified Copernicus Sentinel data [2020], processed by ESA, CC BY-SA 3.0 IGO)

Ioane Teitiota stands outside his home on South Tarawa in the Pacific island nation of Kiribati. The New Zealand Supreme Court turned down his bid to become the world's first 'climate refugee' but his case may open the door to future migrants. (Rimon Rimon)

Leaders of the UN's panel on climate science stand with Monaco's Prince Albert (third from right) at the presentation of a special report on the ocean and melting ice in Monaco. (Alister Doyle)

A view of a narrow alley on the Caribbean island of Gardi Sugdub off Panama. Villagers want to move to the mainland to escape overcrowding on the island, which is also acutely vulnerable to floods and rising seas. (Alister Doyle)

An aerial view of the Caribbean island of Gardi Sugdub off Panama. About 350 metres long, it is home to about 1,450 people living in cramped one-storey homes. (Michael Adams)

Lake Palcacocha, high in the Peruvian Andes, where a melt of glaciers is raising fears of a deadly flood. An outburst flood here in 1941 sent a devastating mudslide down to the city of Huaraz, killing at least 1,800 people. (Alister Doyle)

Ten large plastic pipes siphon water out of Lake Palcacocha in the Peruvian Andes and into the valley below, seeking to limit the risks of a sudden flood from the lake which is being swollen by melting glaciers above. (Alister Doyle)

Peruvian farmer Saúl Luciano Lliuya stands with his lawyer, Roda Verheyen, outside a court in Hamm, Germany. He is suing German utility RWE, saying that its greenhouse gas emissions are melting glaciers in Peru and that the company should pay to prevent floods from Lake Palcacocha above his home. (Alexander Luna)

People board boats in eastern Fiji to visit relatives and friends living on Rabi Island. About 1,000 people were relocated to Rabi Island in 1945 after British phosphate mining ruined their homeland on Banaba Island in what is now the Pacific island state of Kiribati, giving clues to possible migration as seas rise. (Alister Doyle)

A view of Rabi Island, Fiji, where about 1,000 people were relocated in 1945 after British phosphate mining ruined their home on Banaba Island about 2,000km away to the north in the Pacific Ocean. (Alister Doyle)

A stone tidal gauge (foreground) set up in 1837 on the Aaland Islands in the Baltic Sea between Sweden and Finland now has no contact with the water. The land here has been rising relative to the sea since the end of the Ice Age melted a vast weight of ice off the Nordic nations. (Alister Doyle)

An ancient oak tree near Stockholm a few metres above the level of the Baltic Sea. It was used by British scientist Charles Lyell in the 1830s to gauge the rate of land uplift in the region since the Ice Age lifted a vast weight off the land. (Alister Doyle)

This elephant-sized rock at Iggön in eastern Sweden used to be mostly covered by the waters of the Baltic Sea in the sixteenth century and hunters came to spear seals sunbathing on top. The entire coastline has since risen, leaving the rock high and dry in a forest. (Martin Ekman)

climate protests in 2018 by boycotting school in a 'Fridays for Future' movement, a few dozen people are protesting about climate change. Some of them are incongruously holding umbrellas under the blue sky, makeshift parasols against the unusually fierce sun. Thunberg, a global celebrity, is not there today, but her movement has inspired many others, including much older people holding homemade signs reading "writers for future" or "grandparents for future".

With my sister Margaret, who's visiting from England, we meet Ekman for lunch months before the coronavirus pandemic breaks out, in a restaurant in the old town of Stockholm that first opened in 1722, making it a contemporary of Celsius. We have Baltic Sea herrings and discuss the puzzle of the rising land, Celsius and Lyell.

Afterwards my sister and I head to find the oak. It's a magnificent tree, with a vast green canopy, now nowhere near the water line. It has survived the bitter Swedish winters and storms that have left its trunk gnarled with the scars of lost branches. Touching the bark, it's hard to believe this is the same tree as studied by Lyell almost 200 years ago. It's now about 60 metres across a meadow from the water, up a gentle slope. In Lyell's day it was about 40 metres from the water. Further up the slope, a fishing cottage owned by Charles XI of Sweden in the seventeenth century is also still standing in the shadow of another giant oak. For a fishing cottage that used to be by the water, there's more than 100 metres to walk. We measure the girth of Lyell's oak. The 5-metre tape measure my sister has brought along is too short to go right round – we reckon it's 6.4 metres in girth.

Lyell's lecture after his 1834 trip gave away his conclusion that the land was rising, rather than the sea falling, in the title: 'On the proofs of a gradual rising of the land in certain parts of Sweden'. Still, he started off with a nod to Celsius' work. "It is now more than one hundred years since the Swedish naturalist Celsius expressed his opinion that the waters, not only of the Baltic, but of the whole Northern Ocean, were gradually sinking." He went on to note that other scientists were starting to believe that the land itself may be rising. That, he said, sounded outlandish because there were no volcanoes in the Nordic region, nor any history of earthquakes. "It appeared to me improbable that such great effects of subterranean expansion should take place in countries which, like Sweden and Norway, have been remarkably free within the times of history from violent earthquakes. The slow, constant, and insensible elevation of a large tract of land, is a process so different from the sudden rising or falling known to have accompanied, in certain regions, the intermittent action of earthquakes and volcanos, that the fact appeared to require more than an ordinary weight of evidence for its confirmation," he wrote.[7]

Born in 1797, Lyell's life passion was understanding the Earth. Even his honeymoon with his wife, Mary Horner – also a geologist – was spent on a geological tour. As other examples of his diligence in recording sea levels, Lyell noted the height of seaweed washed up on an eleventh-century defensive tower built at Calmar on the south-east of Sweden. He expressed frustration that many in Sweden didn't share his enthusiasm for measuring the rate of change with data, for instance by setting marks in rocks as Celsius had done. "So strong is the conviction of the fishermen here, and of the seafaring inhabitants generally, that a gradual change of level, to the amount of three feet [91cm] or more in a century, is taking place, that they seem to feel no interest whatever in the confirmation of the fact afforded by artificial marks, for they observed to me that they can point out innumerable natural marks in support of the change; and they mentioned this as if it rendered any additional evidence quite superfluous."[8]

Lyell's observations, and sea level marks inscribed on rocks around the Baltic Sea since Celsius' 1731 line, eventually showed conclusively that the rate of sea level change differed along the coast, with fastest rates in the north and least in the south. That meant that the land was rising since a fall in sea level would be the same everywhere. So, he ended up convinced the land was rising, despite his previous doubts. "I am willing … to confess, after reviewing all the statements published previously to my late tour for and against the reality of the change of level in Sweden, that my scepticism appears to have been unwarrantable. ... In regard to the proposition, that the land in certain parts of Sweden is gradually rising, I have no hesitation in assenting to it after my visit."[9]

He said that there were still unknowns – was the land going to keep rising, or might it fall back again? The only way to know would be by constant measurements of the water line. "It is only by multiplying such measurements, and repeating them within short intervals of time, that we shall be able to determine whether the movement of the land be oscillatory or always in one direction, and whether it be intermittent or constant," he concluded.[10]

As Lyell's work in Sweden showed, he was a believer that the processes that originally shaped the Earth were still at work – what was dubbed 'uniformitarianism'. That also implied that changes around the world were happening in slow motion compared to some interpretations of the Bible that the Earth was created a few thousand years ago. Modern science reckons the Earth is about 4.5 billion years old.

Lyell's work was also a boost for his friend Darwin, whose theory of evolution seemed to require millions of years to explain how almost imperceptible rates of change in plants and animals could bring new species with a cascade of natural variations. Darwin published his landmark work *On the Origin of Species* in 1859. "Lyell made time available to Darwin," Richard Fortey, a British palaeontologist, said in an interview published by the Royal Society. Lyell discovered that "the Earth wasn't fixed, but it was capable of up and down motion". Darwin himself once praised "the wonderful superiority of Lyell's manner of treating geology, compared with that of any other author, whose work I had with me or ever afterwards read".[11]

In another sign of more abrupt changes in sea levels in Europe, Lyell's most celebrated work, *Principles of Geology*,[12] shows a picture of the Roman Macellum or 'Temple of Serapis' in Pozzuoli, near Naples, in 1828. Three huge surviving columns at the Macellum are still standing but are scarred by what turned out to be marine molluscs gnawing at the limestone rock – indicating that the entire area had sunk below sea level after Roman times and then rebounded because of the deflation and then ballooning of a magma chamber below ground.

The Lyell Centre,[13] a joint collaboration between the British Geological Survey and Heriot-Watt University, explains Lyell's observations: "Whilst sitting contemplating the site, he observed a line marked by marine *Lithophaga* bivalve molluscs, high up on three columns. He realised that this represented a former shoreline and correctly deduced from this that the site had been submerged for a long period after Roman times and then uplifted again, such that the former shoreline was now approximately 2.74 metres above the present one."

In Sweden, Lyell's observations had helped to show that the land was rising. But even though he had seen evidence of the Earth slowly moving both down and up in Italy, he still could not fathom the processes at work.

In the next breakthrough, Swiss zoologist Louis Agassiz, who became a famed Harvard professor before his belief in white racial superiority and opposition to Darwin's findings tarnished his reputation, correctly suggested[14] in 1837 that the Earth had once had an Ice Age blanketing northern regions. He reasoned that random boulders – made of minerals unknown in the rocks nearby – had been picked up by vast glaciers, for instance from a crumbling mountain cliff, and then carried vast distances entombed in sliding ice. When the Ice Age released its grip, they had been dumped far from their place of origin.

His theory caught on. Scientists now know that global sea levels rose about 120 metres as the last Ice Age – driven by natural shifts in the Earth's orbit around the sun – ended 10,000 years ago and huge ice sheets covering much of North America, Asia and Europe melted into the oceans. Globally, sea levels have been unusually stable in the past 2,000 years,[15] coinciding with the rise of civilisations and cities from Rome to Shanghai. Man-made global warming is now undoing that period of calm.

After Agassiz's findings, it took a few decades to join up the dots, linking the weight of ice sheets during ice ages to the current 'post-glacial rebound' of regions weighed down by ice. In 1865, Thomas Jamieson, a Scottish scientist and friend of Lyell, wrote that there was evidence that the land had been pressed down by the weight of ice. That meant the Earth's crust might not be as solid as commonly thought. "Agassiz considers the ice to have been a mile thick in some parts of America; and everything points to a great thickness in Scandinavia and North Britain. We don't know what is the state of the matter on which the solid crust of the earth reposes. If it is in a state of fusion, a depression might take place from a cause of this kind, and then the melting of the ice would account for the rising of the land, which seems to have followed upon the decrease of the glaciers."[16]

That put the final piece of the puzzle in place. Now, however, global sea level rise is adding another factor destabilising the Baltic. The land is rising, but so now is the sea. That is leading to worries that the Baltic will no longer be a natural bastion against the rising tides.

The pace of global sea level rise – about 3.7mm a year in recent years,[17] and accelerating – means that it is outstripping the natural land uplift in the southern parts of the Baltic Sea and creeping further north as a thaw of Antarctica and Greenland accelerates.

In the northern corner of the Baltic Sea close to the Arctic Circle, the port of Luleå is one of the places least at risk from sea level rise anywhere in the world. Here, the land was once under the thickest blanket of ice during the Ice Age and is rebounding fastest, historically at about 1cm a year.

You can travel 10km inland from the 'New Town' on the coast past a data centre run by Facebook – on land that was under water a few centuries ago – to Luleå's Gammelstad (Old Town). The Old Town is built around a fifteenth-century stone church with more than 400 wooden homes that used to be on an island. The Old Town was an outpost of the then Swedish-Finnish Kingdom near the Arctic Circle. It's now perfectly preserved as a 'church village', on a UNESCO list of world heritage monuments.

Yet even here, Luleå authorities are keeping a wary eye on the long-term trends of climate change, which may mean more severe floods from local rivers. Global sea level rise is also slowing the rate at which the land is rising relative to the Baltic Sea. "The rate of land rise is not 1cm any more, it's more like 7–8mm a year," Lena Bengten, a strategist at the Luleå council and environmental expert, told me in an interview. Bengten comments that Luleå was fortunate compared to many parts of the world, since land rise is a massive cushion against coastal floods.

She was visiting New York in 2012 when Superstorm Sandy swept through the Caribbean and up the east coast of the US, killing a total of 147 people.[18] It also caused US$70 billion in damage in the US alone, with storm surges flooding in the city. "Coming from Luleå it was a shock to see the water flooding into the city," she recalled. As part of long-term planning to protect the region from floods, Bengten said the Luleå council ruled in 2015 that new buildings must be at least 2.5 metres above sea level.

Among drawbacks to the rise of the land and the seabed in the region, the port of Luleå, which exports iron ore, has long had to invest in dredging to scrape away the rising seabed and keep channels accessible to ships. The Swedish government and Luleå port are now investing €310 million[19] in a massive project to deepen the channel to 16.85 metres, about 5 metres deeper than now, to enable bigger ships to dock and make it easier to export iron ore. Sweden's northern Baltic region is the source of 90 per cent of the iron ore in the European Union. Henrik Vuorinen, head of Luleå port, noted that the plans factored in a seabed rise of 7mm a year. The project is likely to be completed in 2027. "In former times, we calculated about 1cm per year but we've decreased it. What happens in 15, 20 or 50 years is still somewhat a guess," he told me. "But we know the [global] sea level is rising."

Further south, some other Swedish municipalities are also acting to limit construction at the shoreline.

The Swedish insurance industry – worried about the extreme long-term risks – is also starting to factor in sea level rise. Insurance companies may in the long term be the main industry driving adaptation to sea level rise – rising premiums to defend against floods in low-lying coastal regions around the world may encourage people to move inland. And coastal property prices may fall, despite the attractions of the waterfront.

"The natural rise of the coast has slowed down, and in the south of Sweden there is no rise at all," said Staffan Moberg, an expert in climate change risks at Insurance Sweden, which groups about fifty insurers accounting for more than 90 per cent of the Swedish insurance market. "But we will see the effects of sea level rise in the future," he told me in a phone interview. Moberg said that coastal cities – such as Gothenburg, Malmo, Stockholm – often have plans to build in low-lying harbour areas, which are also often the historical centres of the city, demolishing old buildings and putting up new ones. "We are trying to say, 'don't build in old harbour areas. We have a lot of land in Sweden. You have to build in safe places instead'. We could eventually see that sea level rises 2 metres, 3 metres, 5 meters over several hundred years. We just don't know," he said. Moberg's argument is that buildings at the shoreline could prove in future to be "monuments to stupidity" even in Sweden.

Despite work in Sweden to understand the interplay between sea levels, geology and greenhouse gas emissions, there has been resistance to the idea of man-made sea level rise. "Sea level is not rising," according to a 2011 essay[20] by Nils-Axel Mörner, who headed the Paleogeophysics & Geodynamics Department at Stockholm University from 1991 to 2005. Mörner, who died in 2020, championed the idea that geological forces like those uncovered by Celsius and Lyell could explain changes in sea levels, not human greenhouse gas emissions. "At most, global average sea level is rising at a rate equivalent to 2–3 inches [5–8 cm] per century. It is probably not rising at all," he wrote, based on visits to places including the Maldives in the Indian Ocean. One of his findings was a small tree growing near the shoreline on a coral atoll – evidence, he said, that sea levels were not rising or it would have drowned from salt years ago. However, the IPCC reckons sea levels rose on average 16cm in the past century, but that the rates vary around the world with geological shifts, or factors such as sand, coral and other deposits shifted around by storms. Some low-lying islands have disappeared, while others have emerged, even as the global trends of sea levels is up.

Mörner flatly disagreed. "Since sea level is not rising, the chief ground of concern at the potential effects of anthropogenic 'global warming' – that millions of shore-dwellers the world over may be displaced as the oceans expand – is baseless," he wrote. "We are facing a very grave, unethical 'sea-level-gate'."[21]

Even a decade ago, such views were at odds with mainstream scientific evidence of the likely risks as seas rise this century. But they struck a chord among doubters of climate change.

The British magazine *The Spectator*, for instance, published a front-page story – 'The Sea Level Scam'[22] – on its front page in 2011, with a cartoon of Mörner examining a palm tree on a tropical island. "The rise and rise of a global scare story", *The Spectator* wrote. Not to be outdone, *The Daily Mail* followed up with the headline[23] "Doomsday predictions on sea level rise are 'false alarm' – levels always fluctuate, says expert as climate change row heats up." Even colleagues at the International Union for Quaternary Research (INQUA), of which Mörner was a prominent member, disowned his research. (The Quaternary is the geological period lasting from 2.6 million years ago to the present.) "You can rest assured that no such scam exists," Professor Roland Gehrels, then of Plymouth University and president of an INQUA commission on coastal change, wrote to *The Spectator*. Gehrels wrote that most of Mörner's claims "have never been published in peer-reviewed articles", that his findings about the Maldives "have all been subsequently refuted" and his claim that the tropical islands were not at risk "is both misleading and dangerous".

A decade on, Gehrels, now at the University of York, told me he was deeply disheartened by Mörner's dismissal of sea level rise.

"I never understood his reasons," he said. Gehrels shared a mail with me that he wrote to Mörner in 2009, saying, "You have always been one of my 'heroes' and my PhD thesis is full of references to your papers". But he added that Mörner's denial of sea level rise meant, "I became concerned that you may be on a political crusade that is now in conflict with the objective scientific facts". Mörner wrote back, gracefully, thanking Gehrels for a "wonderful mail ... I really like honesty. I will consider your words. No, I am in no way on any political crusade – so many are these days – I am quite simply quite convinced of a lack of proper rising evidence."

So almost three centuries after Celsius started scientific research into changing sea levels, the vast weight of scientific opinion is stacked ever more towards alarming risks in coming decades. And there are fewer and fewer climate deniers, like Mörner.

I found documents showing that a group of scientists from INQUA visited Lyell's oak in 1999 as part of a trip along the Swedish coast examining the trend of falling sea levels. The leader of that expedition?[24] Nils-Axel Mörner.

8

ICELAND'S GLACIERS: A REQUIEM FOR ICE

Where the glacier meets the sky, the land ceases to be earthly, and the earth becomes one with the heavens; no sorrows live there anymore, and therefore joy is not necessary; beauty alone reigns there, beyond all demands.

World Light, by Iceland's Halldór Laxness, winner of the 1955 Nobel Prize for Literature

Beloved of poets, the glaciers that put the 'ice' in Iceland are vanishing because of global warming. A memorial in 2019 for one of the disappeared glaciers, Ok, gained worldwide attention while many others are thawing away unnoticed. The nation could be almost glacier-free in 200 years, pushing up world sea levels by about 1cm – and Icelandic scientists have been pioneers in understanding the links between melt and rising oceans.

GROANING AND CRACKING THEIR WAY across the land, with fingers carving deep valleys, glaciers in Iceland are often viewed as vast, mysterious living things, a constant presence on the landscape. But now, laments poet Steinunn Sigurdardóttir, these "emblems of eternity" are dying.

Both sources of terror and wonder since the island was first settled by Vikings in the ninth century, the glaciers are melting into the north Atlantic because of man-made global warming. That is upsetting a land of fiery volcanoes and ice.

Sigurdardóttir spent summers as a girl near Vatnajökull, Iceland's biggest glacier in the south-east, and led cows to pasture past a lava field. "It was the beauty of my life in the summers. It looked like a huge dome that would never go away," she said of the ice. "We have a personal relationship to these beings, these glaciers." Now, by contrast, Vatnajökull is fast "falling apart".

For a nation with 'Ice' in its very name, almost all glaciers on Iceland, about 400 of them, could be gone within 200 years as global warming transforms the inhospitable island that touches the Arctic Circle in the north, Icelandic scientists say.

That has an existential feel for a nation where the hills are alive with the sound of the rivers of ice rumbling downhill, sometimes giving off a metallic smell as layers of buried lava are uncovered in the spring thaw. What might Iceland be called without ice? 'Land'? 'The Country Formerly Known as Iceland'? 'Noiceland'?

Throughout history, Iceland's glaciers have been feared for smothering the landscape, crushing farms, sweeping away roads, bridges and anything else in their way. In one part of the island, volcanic eruptions can cause a catastrophic melt of glaciers and brief cataclysmic floods that can exceed the flow of the Amazon River.

But there is a sense of loss, even mourning, for many at the shrinking of the frozen white giants creeping down from their mountain strongholds.

The knowledge that glaciers are disappearing is becoming ever sharper, not just in Iceland but elsewhere in the world. Both Iceland

and Greenland are experiencing 'last-chance-to-see' tourism in places like Ilulissat in Greenland, where a massive glacier spills spectacularly into a fjord.

Sigurdardóttir said Icelanders still have very differing relationships to glaciers, from fondness to aversion. She spoke to people living near the ice about their childhood memories. "Some even said they were taken by their grandmother to pet the glacier, like a dog or a cat. In other farms where the glacier caused flooding they were told 'don't go near to the glacier'. We have a lot of mixed feelings here. Nature in Iceland has been so harsh. We are tainted by the fact that living in the country was like living in hell for many centuries. There was one volcanic eruption after another, one plague after the other, and people were totally poor," she said. Ash could destroy crops, choke animals.

Helgi Björnsson, a professor emeritus at the University of Iceland's Institute of Earth Sciences and a leading global expert, says the nation's glaciers are doomed as temperatures rise, unless there is some radical action to halt climate change by cutting greenhouse gas emissions.

Glaciers cover about 10,000 sq km of Iceland, roughly 10 per cent of the land, he tells me on a Zoom call. This century, the area of Iceland's glaciers has reduced by more than 800 sq km, according to a joint study by the University of Iceland, the Icelandic Meteorological Office and the Southeast Iceland Nature Centre.[1] The study also found that they have shrunk by 2,200 sq km from the end of the nineteenth century when the glaciers reached their maximum extent since the country was settled in the ninth century.

Glaciers grew during what's known as the Little Ice Age,[2] a natural cool period lasting from about 1550 to 1850 and affecting mainly Europe, North America and Asia, before man-made greenhouse gases took over as the main driver. The Little Ice Age coincided with high levels of volcanic activity, throwing sun-dimming ash into the air.

Björnsson gives me a depressing timetable to doom.

Glaciers are on average 350 metres thick and, with global warming, they are losing about 1 metre of thickness every year as temperatures rise. That implies a lifetime of less than 350 years on current trends before Iceland is ice-free, because the shrinkage will accelerate over time. And some small glaciers have already thawed away, lost in remote mountain regions before anyone noticed. "We have already lost about 50 glaciers," he tells me.

A few glaciers, luckily, will be far more resilient: Vatnajökull, the biggest and covering about 7,700 sq km,[3] is 1,000 metres thick and so less vulnerable to the global thaw. But there are many factors at work, beyond greenhouse gases. In some places, as the ice melts on mountaintops, the highest part of the glacier will get lower, where air temperatures are warmer, accelerating the losses. In other places, there may be more rain in the summertime, reducing the ice.

"For the glaciers, 200 years from now only some parts of Vatnajökull remain," Björnsson said of the projected impacts of global warming. The loss of all the nation's glaciers would push up global sea levels by about 1cm, tiny compared to Greenland whose ice sheet is equivalent to 7 metres of sea level rise.

"It is in our blood, our identity will be damaged if the glaciers disappear," he said. "Almost every poem about Iceland includes something about the glaciers."

Glaciers build up from snow falling in winter that gets compressed into ice. Over many years, as ever more snow and ice amass on a mountainside, the glacier starts to flow downhill, under the force of gravity. To grow, glaciers need the build-up of snow and ice to exceed the summer melt. But glaciers shrink, and can eventually die, when the summer melt outstrips the supply of new snow.

Svanhvít Jóhannsdóttir, aged 27, whose first name translates into English as 'White Swan', works as a park ranger and lives by a finger of the Vatnajökull glacier in south Iceland. On a video call in spring 2021, she points the camera out the window over the hillside where she used to be able to see the ice from her house. It has receded up the hill and out of view. She says there's a sharp difference between younger and older people in their attitudes towards glaciers. "For the younger generation, like myself, I have always grown up in the discussion of climate change. For us, it's a tragedy and a serious matter, sad and hopeless. There is such big evidence from what we read, and what the science is telling us." But she said that her parents' generation "have lived a totally different experience towards glaciers". Just a few decades ago, "the glaciers were just a big hazard ... People died on the glaciers trying to get across or trying to deal with them."

She noted that Iceland only managed to complete a 1,300km ring road around the island, Route 1, in 1974 after the construction of a 904-metre-long bridge across the Skeidará River in the south-east. The bridge covers a treacherous sandy plain where rivers wash out debris and water from the Vatnajökull glacier system. Before the bridge was built, people had to resort in olden times to travelling inland and hiking across the glacier when the river was impassable. More recently, they could take a short-hop flight from a gravel airstrip or spend ages driving around the whole nation. "It was quite an achievement that they managed to tame the river. I can understand that those people who had those interactions with the glacier... find it quite a relief that they are getting smaller," she reflects.

Jóhannsdóttir did a thesis as part of her degree at the University of Akureyri titled 'The impact of glaciers on people: global, cultural, social and cognitive'. She interviewed many Icelanders for their thoughts about ice. "There was a lot of mention of the glacier as a person, it's breathing, it's crawling, or dying. Words they would use to describe something alive. I think it's natural that people would speak of some kind of living beast that had maybe a heartbeat, or that it breathes," she comments. For some people, moulins – holes in the ice like wells where water flows into the glacier – "are somewhat similar to a vein system in a body". Young people also saw opportunities from glaciers – they generate jobs from tourism for guides, restaurants and hotels.

Even so, glaciers may not be crucial for Iceland's identity. Many Icelanders take pride in their survival in a land of fire and ice, where standing up to the elements is a sign of national strength. "Icelanders' identity is based on the harsh environment, the harsh weather, and the glaciers are part of it. But they are not all of it," she said of her conclusions. "Even if the glaciers would disappear there would still be the element of harshness in nature – volcanic activity, steep mountains and bad weather. It's hard to say that the glaciers play any kind of dominant role in Icelanders' identity. But they are part of it."

And even young people are not unified in their devotion to glaciers. Jóhannsdóttir reflects, "I sometimes forget that I am in a bubble out here in the countryside, surrounded by my friends who are in a similar profession. The reality is that there are also a lot of young people, maybe in the city, who are not thinking about these things and maybe do not care so much. So there's also variety within the age groups."

In 2014, a small glacier led to a big international stir.

That year, a local newspaper, *Morgunbladid*, and Icelandic television reported on the disappearance of a small glacier, near Reykjavik, known as Okjökull, or Ok glacier. In Icelandic, 'ok', pronounced 'awk' means a yoke for an animal, or a burden, and 'jökull' is the word for 'glacier'.

This ex-glacier was to go on to win international fame – as 'Not OK'. *Morgunbladid* quoted glaciologist Oddur Sigurdsson as saying Ok was "dead ice" and no longer met the criteria for being a glacier as that requires it to move. In 1890, Ok had covered 16 sq km, but has been steadily declining ever since and was now too small, he said.

"It had been dead ice for quite some time before 2014," Sigurdsson told me by phone. "Somewhere around the turn of the century it stopped being a glacier." Many glaciers in Iceland don't have names – Ok was the first named glacier to thaw away. I found him by checking in the Icelandic phone directory – my heart sank when it listed about ten people with the same name but luckily he was only the second person I dialled. Sigurdsson said that when snow and ice becomes about 40–50 metres thick, the weight transforms the ice into a material that can flow, and so qualifies as a glacier. Any less, and the glacier dies. He told me that he was currently compiling an updated list of glaciers and ex-glaciers in Iceland. "Ok is by far the largest glacier to have died," he said. It is also near Reykjavik, the capital, and so was better known than many of the more remote rivers of ice that have stopped flowing.

Referring to others on the verge of expiring, he mentioned Hofsjökull, a tiny glacier in the south-east that shares its name with Iceland's third-biggest glacier. The small Hofsjökull had an area of about 4 sq km in a 2008 overview[4] by Sigurdsson.

"It will be gone in the next decade, if it is not already expired as a glacier ... it has signs of death," he said.

Issuing 'death certificates' for an inanimate lump of ice may sound a weird idea and the passing of Ok remained a small item in Icelandic news until a couple of US anthropologists, Cymene Howe and Dominic Boyer of Rice University, stumbled across a news item about Sigurdsson and Ok on an English-language website: Icelandic Review. It was less than 100 words long. "It was a kind of a shrug. Very little fanfare, no feature story, no reckoning with what this means," Howe told me on a Zoom call. For many people, it was merely "a small glacier that died in a faraway place that nobody cared about. Not even Icelanders."

But to her and Boyer, "this seemed like a big deal", she said. Their project also involved speaking with Icelanders about their relationship to ice. Like Jóhannsdóttir's work, it underscored how many people thought of glaciers as being alive. "Glaciers have snouts, they have tongues, they have toes," she said, referring to terms for the end of glaciers. "To be alive they have to move under their own weight. They seem to have their own volution, even though they are inanimate objects. Once they stop moving, they become dead ice. Yet they were never alive, but they are animated." She noted how views of glaciers have changed through history. *The*

Book of the Settlement of Iceland,[5] written in medieval times, says that Iceland was first discovered by Irish Christians who left "Irish books, bells and croziers". These were found by Vikings from Norway who were the first permanent settlers of the island about 1,200 years ago. The book names Ingólfr Arnarson as the first permanent settler, in 874. It also says Iceland got its name from a Norwegian settler, Flóki Vilgerdarson. When he landed there, "The bay so abounded in fish, that by reason of the catch thereof they gave no heed to the gathering in of hay, so that all the livestock perished in the winter. The following spring was rather cold; then Flóki went up to the top of a high mountain and discovered north, beyond the mountain, a firth [coastal inlet] full of drift ice; therefore they called the land 'Iceland', and so it has been called since then."

By contrast, Greenland, which is far icier, got its name from Viking adventurer Erik the Red a century later in what may have been a more successful bit of Viking propaganda. The saga says bluntly that he reckoned the name Greenland "would encourage people to go there that the land had a good name".[6]

Unlike the apparent attractiveness of Greenland, the first reports back to Scandinavia about Iceland were contradictory. Floki "spoke ill of it" but one of his fellow settlers said, "butter dropped from every blade of grass in the land which they had discovered", according to *The Book of the Settlement.*

About 1,000 years ago, both Greenland and Iceland enjoyed a Medieval Warm Period that may have made the region more habitable. Greenland's European colonies mysteriously disappeared by the 1460s, perhaps as the region cooled. In the early nineteenth century, Howe at Rice University said there was a shift in Europe, including Iceland, towards appreciating "the pastoral, romantic feeling of being with nature" and away from the suspicions of the ice. This was a time when Romantic poets including Byron were also writing about the ice, far away in the Alps – he described visiting a glacier above Grindelwald in Switzerland: "like a frozen hurricane; starlight beautiful, but a devil of a path".[7]

That strand of appreciating the beauty of glaciers has since grown, accelerated in recent decades by the widening awareness that they are disappearing because of global warming driven by man-made emissions of carbon dioxide.

Jón Yngvi Jóhannsson, a professor of education at the University of Iceland, also said there was a shift in Iceland since it won full independence from Denmark after a wartime referendum in 1944. "From the beginning of the struggle for national independence until the middle of the twentieth century, the focus in everything that had to do with national identity was always on history, on literature, on language," he told Howe in a documentary about Ok. Now, appreciation of the beauty of Iceland's nature was getting a bigger role.

In recent years, Iceland, which suffered a massive financial crisis in 2008–11 when its three main banks defaulted and the Icelandic crown plunged in value, has become more affordable for foreign tourists intrigued by ice. So, the Rice University scientists and colleagues struck on the idea of holding a formal memorial for Ok in 2019, and installing a plaque at the top of the mountain.

"We kept hearing this narrative of loss and death," Howe said of interviews with Icelanders. "A memorial is something that is emotional, even if it is for something that never lived," she said. "Everyone knows what it is to lose someone. So there was something universal" about the idea of a ritual to grieve for the ice.

For the 18 August 2019 memorial, the scientists expected about twenty or so people to turn up – more than 100 came. And the death of the small glacier reverberated around the world in countless news outlets as a symbol of the need for more action to slow climate change. Icelandic Prime Minister Katrin Jakobsdóttir was among the 'mourners' who came to the base of the mountain, along with UN High Commissioner for Human Rights Mary Robinson.

"I hope that this ceremony will be an inspiration, not only to us here in Iceland, but also for the rest of the world because what we are seeing here is just one face of the climate crisis," Jakobsdóttir said, at the base of Ok. She didn't have time to make the climb, but, wearing bright-coloured jackets and snow gear against the chill winds, the mourners clambered in single file over a desolate mountainside of boulders and ice up to Ok's peak at 1,198 metres.

People carried small banners: 'RIP Ok glacier. Not OK!', 'Declare climate emergency now!', 'Global climate crisis' and 'Pull the emergency brake'. Some sucked on bits of ice for a final taste of the departing glacier.

At the top, they inserted a plaque in holes drilled into a large rock, with the text in Icelandic and in English. Written by poet, novelist and playwright Andri Snaer Magnason, it reads: "A letter to the future. Ok is the first Icelandic glacier to lose its status as a glacier. In the next 200 years all our glaciers are expected to follow the same path. This monument is to acknowledge that we know what is happening and what needs to be done. Only you know if we did it." The plaque ends with the concentration of carbon dioxide in the atmosphere in August 2019 – 415 parts per million (ppm). That is already a sharp gain from about 280 ppm for thousands of years before the Industrial Revolution.

If governments manage to halt and reverse global warming, there is of course some hope that the plaque will be covered up, entombed in a new glacier. But it's likely to stay visible, apart from coverings during winter blizzards, as the melt widens across the island.

At the time, Magnason wrote in *The Guardian*: "According to current trends, all glaciers in Iceland will disappear in the next 200 years. So the plaque for Ok could be the first of 400 in Iceland alone. The glacier Snæfellsjökull, where Jules Verne began his *Journey to the Centre of the Earth*, is likely to be gone in the next 30 years and that will be a significant loss. This glacier is for Iceland what Fuji is for Japan."[8]

Jules Verne's glacier is to the west of Ok, and it marks the spot where travellers descend below ground in the 1864 novel.

Magnason also noted that the year 2100 – typically the year used for projections of climate change – sounds like a long time away.

But many people born today are likely to be alive in 2100, and their grandchildren in the latter part of the next century.

Howe and Boyer made a short film, *Not Ok*,[9] with former Reykjavik Mayor Jon Gnarr pretending to be the voice of the mountain itself. He growls slowly like a cantankerous giant regretting how he had been overlooked in Viking records in favour of more scenic mountains nearby. "This is me. They call me Ok. I'm a mountain. Well, really a volcano, or at least a former volcano. I've been around here for as long as there has been here to be around. And for most of that time I've had a glacier on my back," he intones. Howe said his deep voice reverberates even better because it is slowed to 80 per cent of normal speed.

Among coverage of this glacier's passing, the British newsweekly magazine *The Economist* broke with its usual editorial and philosophical norms about the meaning of life by devoting an article to Ok on its obituary page.[10] Its tribute about "the end of ice" quoted a high school pupil as reading a poem to "Ok, the burdened glacier / which at last had had enough / of acts of terror from men who do not know / how to have both profits and morals".

Howe says it is hard to pin down exactly why the memorial to Ok touched a worldwide nerve when many other glaciers are also vanishing. Partly it was topical, with a focus both on climate change and the Atlantic. And the slogan 'Not OK' resonated.

Thunberg, who does not travel by plane to limit her greenhouse gas emissions, was at the time on a racing yacht crossing the Atlantic from Britain to New York to attend a UN climate summit. She had left Plymouth in England on 14 August for the two-week trip, amid wide media coverage of the bare-bones *Malizia II* vessel – no kitchen, no shower, no toilet – and a sail emblazoned '#FridaysforFuture'.

Climate scientist Martin Stendel of the Danish Meteorological Institute had also drawn attention to melting ice in the region by tweeting[11] in early August that a record 12 billion tonnes of ice had poured from Greenland in just two days. He said that would be "enough to cover Florida with almost five inches of water ... or in metric units Germany with almost 7cm of water ... or Denmark with half a meter of water".

Prime Minister Jakobsdóttir wrote an article in the *New York Times*[12] drawing attention to the ceremony on the eve of the trek up to Ok, urging global action to reduce emissions. "Ok is no longer a glacier," she wrote. "The ice field that covered the mountain in 1900 – close to six square miles – has now been replaced by a crater lake. It is certainly beautiful, surrounded by patchy snowfields, and is now the highest lake in Iceland. But that beauty quickly fades in the eyes of anyone who knows what was there before and why it is no longer there. Ok's disappearance is yet another testimony of irreversible global climate change."

"This is a local ceremony but a global story," she concluded. "We know what is happening and what needs to be done. Help us keep the ice in Iceland."

And Howe said Icelandic authorities helped by giving quick approval for the plaque to be drilled into a rock on the mountainside. "Icelanders like quirky stuff too," she said.

Even so, the attention paid to Ok's demise is hard to fathom. It might even have happened decades ago. Björnsson sent me a paper he wrote in the Icelandic journal *Jökull* in 1979,[13] detailing glaciers over many thousands of years. It says a general recession of glaciers set in during the 1890s as the Little Ice Age came to an end. It includes the lines: "Districts cultivated by farmers in the twelfth century are exposed again. The Ice caps on Ok and the Glama plateau have disappeared."

Glaciers are dying elsewhere too, pouring billions of tonnes of water into the oceans. In the Andes, for instance, the Chacaltaya glacier near La Paz in Bolivia – once known as the world's highest ski resort at 5,300 metres, with a single lift – was pronounced expired more than a decade ago by Edson Ramirez, head of an international team that had been studying the glacier.

As in Iceland, the Bolivians saw the glacier as a living thing. "The glacier is dead. Chacaltaya is dead," Ramirez said in a UN broadcast.[14] "It's very sad to find that the glaciers are actually disappearing. It's very dramatic, very, very dramatic." A BBC report covered how he organised a ceremony to commemorate the glacier's demise. "It is as if someone died," he said.[15]

And more and more glaciers are 'dying', though most studies avoid the word for drier academic terms.

"The melt rate and cumulative loss in glacier thickness continues to be extraordinary," the World Glacier Monitoring Service (WGMS) wrote in a 2020 report.[16] It said that the loss of glaciers amounts to about 335 billion tonnes of lost ice each year, almost 1 millimetre of sea level rise every year. Some glaciers expand in some years, for instance when there is unusually high snowfall, or a chilly summer against the global warming trend. But a WGMS overview of reference glaciers[17] around the world, starting in 1950, shows that glaciers overall gained mass only in five years, with no net year of gain since the 1980s.

Along with mourning the passing of glaciers, Icelanders are also pragmatic. It's not all about poetry and mourning, but also about economics. Among spinoffs, the extra water flowing from glaciers, stoked by burning fossil fuels, is helping to raise hydropower generation.

Iceland gets its electricity from renewable sources[18] – with about 73 per cent from hydropower, 27 per cent from geothermal, and a tiny 0.01 per cent from wind. It is expanding hydropower generation to take advantage of the melting glaciers.

"The flows are increasing, they will continue to increase until the middle of the century. After that they will subside very slowly," said Óli Grétar Blöndal Sveinsson, Vice-President of Hydropower Research and Development at national power company Landsvirkjun. New hydropower turbines can take advantage of the extra flows that would otherwise just spill over dams and run into the sea. When the glaciers disappear, Iceland will have to rely on water from rain and melting snow flowing into reservoirs. He said he counted the expanding flow from hydropower as 'renewable energy' even though it is a one-off melt, but added, "You can put a question mark – the increase itself from the glaciers is not renewable".

Among other projects, Landsvirkjun added an extra 100MW installed capacity to the Burfell[19] Power Station in 2018, raising capacity to 370MW from 270MW in a US$160 million project. Sveinsson said the expansion did not seem to make economic sense when originally proposed a few years ago – but it did when factoring in increased flows from glacial melt prompted by global warming. One problem limiting new investments in hydropower was that Iceland used to use climate and meteorological records dating back fifty years, a period that includes times when glaciers grew. But Landsvirkjun switched to only considering the most recent twenty years, when the warming trend is a one-way signal of thaw.

Always lurking in the background for Icelanders is the fear of the next big eruption of its volcanoes. In March 2021, the Fagradalsfjall volcano erupted near Reykjavik, for the first time in about 800 years. It is not viewed as a major threat. By contrast, in 2010, the eruption of Eyjafjallajökull, Iceland's sixth-largest volcano, caused widespread air travel disruptions as flights were grounded because of fears the ash would clog up engines.

Iceland has far bigger threats, including *jökulhlaups*, a type of outburst flood from a lake beneath a glacier. "No glacier poses a greater threat to Iceland than Mýrdalsjökull, for beneath it lies hidden the country's most infamous volcano, Katla, which has burst through its glacial cover twenty times since the settlement of Iceland, causing enormous, rampant jökulhlaups, that have swept over gravel plains, roads, bridges, farmsteads, and cultivated land," Björnsson wrote in his book, *The Glaciers of Iceland*.[20]

The last major eruption by Katla was in 1918, and the century-long pause is getting worryingly long. "The big outburst floods from Katla are the largest floods on earth," Björnsson told me. "We may have had Noah's flood here." An eruption in 1755 unleashed a flow of water equivalent to 100,000 to 300,000 cubic metres a second for a few hours – a cataclysm enough to cause tsunamis out at sea, he said. That makes Katla, briefly, comparable to the Amazon, by far the world's biggest river in terms of the water flow. The Amazon discharges about 200,000 cubic metres per second[21] into the ocean off Brazil.

Poets have long struggled to put the horror of glaciers and eruptions into words. Björnsson said some believe that 'The Prophecy of the Seeress', part of *The Poetic Edda*, thought to have been written in the late tenth century, may refer to Katla: "The sun turns sable, land sinks under sea, banished from the sky are clear bright stars. Steam bursts forth with flashes of fire, burning hot flames flicker at heaven itself."

Among other disasters, a massive 1783 Laki lava eruption of the Grimsvötn volcanic system in southern Iceland cast a shade of sulphuric acid over the northern hemisphere that pushed down global temperatures. It caused crop failures from Britain to Japan and some researchers suggest the suffering and famine may have been among the contributing factors to the French Revolution of 1789.

Benjamin Franklin, one of the US Founding Fathers, who was chief US diplomatic representative in France at the time, wrote in a letter[22] in 1784: "There existed a constant fog over all Europe and great part of North America. This fog was of a permanent nature; it was dry, and the rays of the sun seemed to have little effect towards dissipating it." He was among those who speculated that it was caused by Iceland's volcanoes.

There are also fears that volcanoes entombed beneath the pressure of hundreds of metres of ice could awaken as the ice melts.[23] Less pressure on magma beneath as the vast lid of hundreds of metres of ice thins could release more volcanic activity, Björnsson said. Among worrying signs, the ending of the last Ice Age about 10,000 years ago coincided with a surge in volcanic activity in Iceland, a sign of how fire and ice can form a dangerous duo.

Nowadays, for some scientists, eruptions may give a blueprint for fixing global warming by creating tame versions of Katla, Grimsvötn, Eyjafjallajökull and other volcanoes. The idea is that scientists could deliberately spray sulphur-rich particles into the upper atmosphere, lifted by balloons or planes, to form a planetary sunshade. Supporters of such geoengineering say that could give us control of a crude global thermostat, and even allow us to cool the planet. Who knows: at best a cooling might even allow Ok to revive, like a Lazarus glacier. But such climate engineering comes with huge and hard-to-quantify risks, including that it could shift global rainfall patterns or discourage nations from making needed cuts in greenhouse gas emissions.

It would also mean living with more air pollution and would not solve the build-up of carbon dioxide in the atmosphere which is acidifying the oceans and harming marine life from corals to lobsters. So, many nations are deeply sceptical. In 2010, governments agreed a moratorium[24] on geoengineering activities under the UN Convention on Biological Diversity. It allows an exemption for small scale-scientific research.

For rising seas, an Icelandic scientist was one of the pioneers in linking the water pouring off the land to the level of the oceans. Sigurdur Thórarinsson, who lived from 1912 to 1983, was a geologist whose skills included writing songs that made it to the top ten on Reykjavik radio and training astronauts for the Apollo moon project. He taught astronauts to recognise and collect interesting rocks in the 1960s, when barren Iceland was seen as the closest thing to the moon on Earth.

He grew up in a remote part of the island which had changed little over the centuries and where travel was often on horseback. “I sometimes say I grew up in the Viking age and ended up training the astronauts,” Thórarinsson, wearing glasses with thick black rims, said in an interview[25] for the University of Lund in 1979. He said he was sometimes too little of a specialist. “I sometimes regret that I may be a little too broad and a little too shallow,” he added. Thórarinsson also had a good sense of dark humour. When asked when Katla, the destructive volcano, would next erupt, he said he could not say exactly, “but it is my sentiment that I shall see the next eruption from beneath”.

He studied glaciers in Iceland and around the world to learn about sea levels. "The majority of the glaciers in practically every glacier district of the world are now receding, i.e. the present glacier shrinkage is a universal phenomenon," he wrote in 1940. "At present one must reckon with the sea-level being raised a minimum of 0.3mm per annum in consequence of the present glacier shrinkage", he wrote in a 1940 article[26] – one of the very first to link melting ice to rising sea levels. He did not speculate about future rates, or the causes, but urged more study. That, he concluded, would "open yet another path leading towards the solution of the difficult and complicated problem ... [of] how the climate has developed".

Thórarinsson seems to have been a charming man – someone to cheer you up in the harshness of the Icelandic climate. "He wrote many songs or ballads to entertain his co-workers and friends in the Icelandic glaciological society, when they got stuck in bad weather during their annual excursions to the Vatnajökull glacier," his son Sven Sigurdsson told me in an email. "Most of these ballads remained private but a few became public and some of them remain popular even today, although no doubt Sigurdur would have liked to be remembered longer for his scientific achievements rather than his songs!"

Björnsson wrote a fond obituary to Thórarinsson, calling him the "Grand Old Man of Geoscience" in Iceland who was "blessed with more volcanic eruptions than any other Icelandic geologist".[27] Björnsson told me that Thórarinsson had been at a conference in the Himalayas in 1960s. "The weather became very bad ... what saved them was Sigurdur. He started singing. It was easier for them to stay there in isolation with him keeping up their spirits. They were quite surprised that this eminent professor could do this". So, having fun and surviving in hostile conditions is part of life in Iceland, a source of national pride.

The realisation that mighty glaciers are fragile has happened abruptly in recent years, as greenhouse gas emissions have driven up world temperatures. Sigurdardóttir, the poet, said she noticed she treated the glacier "as a symbol of eternity, something that will always be there" in a 1995 novel, *Place of the Heart*. And yet just ten years later, she described a woman dying out in the countryside. "While she is dying she is watching the glacier and is thinking that everything is transient, even the glacier is not going to last," she said of her novel *Sunshine Horse*. She also wrote a book of poems as an elegy for the dying Vatnajökull glacier, *Dimmumót/Nightfall*, published in 2019. Among the poems inspired by her childhood are 'The Daydream Trail':

> On the old trail I dreamt dreams of days
> Driving the cows, two of them, and two calves, onward, past a mossland
> now arid grey, now rainwashed green.
> Lagging beasts ruminating – in time with the slow-paced dream.
> Over our radiant, lackadaisical progress a gleaming ice-mountain kept watch.
> A vault reaching to heaven wreathed in fluttering clouds.
> And the sun came briskly to life.
> Here the white mountain shines on its lover, the sun, the body of heaven.
> Thus would the time of sunshine, the time of imminent Beauty, come to be called Life Itself.
> Which had not begun.
> As the child knew, just then, 'twas but a dream, and
> Reality would follow, so much greater and more exalted than what was just dreamt.

9

FIGHTING SEA LEVEL RISE: BUILD A WALL, OR MAYBE A SANDCASTLE?

Life on this earth first emerged from the sea. And as the polar ice melts and sea level rises, we humans find ourselves facing the prospect that, once again, we may quite literally become ocean.

John Luther Adams, American composer

What to do about rising sea levels? Some scientists are floating radical ideas such as building sea walls between Britain and Europe to seal off the North Sea from the rising Atlantic Ocean, or pumping vast amounts of sea water onto Antarctica where it would refreeze. For now, planners in places from the Netherlands to Florida are focused on immediate shoreline problems, using dykes or vast amounts of dredged-up sand to defend their coasts.

A DUTCH SCIENTIST HAS A radical plan[1] to combat sea level rise. He hopes no one will ever use it.

The Netherlands has long been a world leader in protecting its citizens from inundation with dykes since 26 per cent of the land – homes, business, railways, roads, farms and airports – is below the level of the North Sea. A suggestion by Sjoerd Groeskamp, an oceanographer at the Royal Netherlands Institute for Sea Research, goes far further than any current policies. He proposes mega-dams – far bigger than any ever built – to seal off the North Sea, the Baltic Sea and the English Channel into a single new sea if nations fail to rein in their greenhouse gas emissions. The dams would cut the new sea off from the Atlantic Ocean, which would keep rising as trillions of tonnes of ice melt from Antarctica to Greenland. The northern section of the proposed dams, 476km long, would run from the north of Scotland, via the Orkney and Shetland Islands, to Bergen on the west coast of Norway. A shorter section, 161km long, would connect Cornwall in south-west England to Brittany in western France. Inside the new artificial sea, the level of the waters lapping the coasts of much of Europe would stay unchanged, except for the inflow from rivers.

Groeskamp and his co-author of the project, Joakim Kjellsson of the Heimholtz Centre for Ocean Research in Kiel, reckon the dams would cost between €250 and €550 billion, or about 0.1 per cent of the annual gross domestic product (GDP) of the fifteen countries involved, for each of the twenty years it would take to build. Even that massive bill would, they reckon, be far less than the costs for each nation separately to keep protecting thousands of kilometres of coastlines – with walls or other measures to protect ports, roads, nuclear power stations and farmland – if sea level rise keeps accelerating this century. And it would be much cheaper than the damage wreaked by floods if nothing was done.

Deep cuts in greenhouse gas emissions are the best way to limit global warming. But bold action has so far failed to live up to hopes generated by the 2015 Paris Agreement and by ever more national pledges to work for net zero emissions by mid-century.

So far, almost all research into muting the impact of sea level rise focuses on national or regional plans to protect coastlines, especially cities that have the most valuable real estate in most nations – from Amsterdam to New York, Jakarta to Guangzhou. But worries that the world may be reaching a point of no return in the thaw of Greenland and parts of Antarctica are spurring ever more drastic ideas.

When I talk to Groeskamp on a video call, I start by saying, "My first reaction was that your project's crazy." I'm a bit worried that he'll be insulted, but he takes it in his stride, "That was my first reaction too ... But if you compare it to the alternatives, it's not as crazy as I thought."

The proposed dams would protect fifteen nations – Belgium, Denmark, England, Estonia, Finland, France, Germany, Latvia, Lithuania, Poland, the Netherlands, Norway, Russia, Scotland and Sweden. More than anything, Groeskamp says the dams are a warning about the agonising choices nations may have to take if they fail to act and the oceans rise by metres in coming centuries. Few governments, so far, are looking at risks of sea level rise beyond 2100 – even though that is within the lifetime of many children born today.

As we have already explored, the IPCC reckons that ocean levels could rise by a metre or so by 2100, and up to about 7 metres by 2300 in the worst likely case. And it will be ever harder to ignore: on the high trajectory, seas will be rising by more than 1.5cm a year in 2100.

The suggested sea dams would have massive drawbacks. They would disrupt international shipping as vessels would have to go through locks to get back and forth to the Atlantic. The new sea behind the dams, from the western English Channel to the Baltic Sea in Finland and the North Sea up to Norway, would be more stagnant without the swirl of currents from the Atlantic, and perhaps ever less salty as rivers pour in fresh water. That would have untold harm for marine life – everything from cod to mackerel, from seaweeds to sharks.

Plus, how would Russia, for instance, feel about dams connecting NATO nations Britain, France and Norway that could block its access to the Atlantic from Kaliningrad and St Petersburg on the Baltic Sea? In Britain, would Wales pay for a project that only defends parts of England and Scotland? How about the risk of terrorist attacks that could blow holes in the dam?

The study, published in the *Bulletin of the American Meteorological Society* in August 2020, does not go into the details of the possible impacts on geopolitics or ecology in its overriding goal of highlighting the risks of inaction on climate change.

Still, in purely economic terms, Groeskamp said the project would make sense. According to him, in the Netherlands alone, ensuring coastal protection against 1.5 metres of sea level rise by 2100 would be about a third of the total cost of building the dams. The Dutch now spend about €1 billion a year on 'dry-feet insurance' – building and maintaining protection against the waves.

The study, widely covered by media around the world, cuts to the heart of wrangling about how to protect coasts. Do we keep building higher protection for coasts, or do we move people inland in a managed retreat, or simply abandon the shores? "We don't want these dams. We want to avoid this. But we have to be prepared for the worst-case scenarios", is Groeskamp's thinking.

On social media, some were angry at the consequences of choking the seas, he said. Others were more joking: one tweet showed a dam around Britain, isolating it from the rest of Europe to underscore its departure from the European Union under Brexit. One commentator called Geofbob wrote, "Suddenly Elon Musk's plan of us all emigrating to Mars doesn't look so crazy." Surfing enthusiasts shared worries that the new sea would be like a millpond deprived of waves rolling in from the Atlantic.

Maybe the plan touched a nerve because dams are easily imagined as a starting point for a debate about the benefits and risks of limiting rising seas. It's also easier to grasp than more abstract things like rising temperatures, carbon dioxide concentrations in the atmosphere or the pace of sea level rise.

Groeskamp said he took most of the attention as a compliment – it showed it had stoked debate. But he was saddened by some ecologists who slammed the idea, saying, "This will kill the ecosystems". Critics "didn't see the whole notion that it was a warning. My reaction is 'use this in your favour' because this is what will happen if we don't do anything about climate change," he said. Marine biologists, for instance, could use the fear of dams to put pressure on governments to act now, rather than dither and risk running out of sensible options. Every year of waiting in cutting greenhouse gas emissions adds to the harms caused by global warming. Other reactions that troubled him were that building massive dams could be an excuse for inaction – people might shrug and say, 'Just build a giant dam and that will fix climate change'. That, he said, was the opposite of what he intended by warning that time is running out. His study is titled 'The Northern European Enclosure Dam for if Climate Change Mitigation Fails', but Groeskamp said he originally wrote a more fatalistic title, 'When Climate Change Mitigation Fails'. His editor softened 'when' to 'if'.

The idea of damming a sea is not without precedent. In South Korea, the 33.9km-long Saemangeum Seawall cost €1.8 billion and has an average height of 36 metres above the seabed and is 290 metres wide.[2] The wall was completed in 2010 and separates the Yellow Sea from the former Saemangeum estuary, to create 400 sq km of land for industry, tourism and farming. The project has been bitterly criticised by environmental campaigners who say it ruins coastal wetlands vital for migratory birds and other wildlife.

The wall surpassed the 32km Dutch Afsluitdijk, completed in 1932, as the world's longest sea wall. The Afsluitdijk dams the Zuiderzee, a salt-water inlet of the North Sea created by erosion in the twelfth and thirteenth centuries. The dam turned the Zuiderzee back into freshwater, Ijsselmeer.

Even so, the proposed northern European dams are on a far bigger scale.

In the north, it would have to fill in the Norwegian Trench, which has a maximum depth of 321 metres. All together, the project would require 51 billion tonnes of sand just to make the concrete, roughly equal to a year's global consumption of sand. And enormous pumps would be needed on the dams to spew water into the Atlantic. Otherwise, the new isolated sea would slowly fill up with water flowing in from rivers. An estimated 40,000 cubic metres of water every second would pour in from rivers including the Rhine, the Seine and the Thames. Among the world's biggest pumps is a pumping station in New Orleans, built to combat storm surges and floods caused by hurricanes, and it has a capacity of about 540 cubic metres per second, according to the US Army Corps of Engineers.[3] So, the proposed European dams would need more than seventy pumping stations with that capacity, able to operate year round.

Groeskamp reckons that other seas might be dammed – the Mediterranean could be isolated by blocking the 13km wide Strait of Gibraltar between Spain and Morocco. That idea has been around for a long time – a century ago, a German architect proposed blocking the strait to lower the water level in the Mediterranean and create new land. Unsurprisingly, it was never built.

Other candidates for closure with vast dams could be the Red Sea, the Persian Gulf or the Sea of Japan. There would also be enormous environmental risks. The Mediterranean, for instance, feeds salty water into the Atlantic that helps drive the Gulf Stream, a vital source of tropical water flowing northwards that keeps Europe relatively warm.

Plus, Groeskamp's project would only address sea level rise for part of Europe, not the world. Europe's cities are not as vulnerable as many others elsewhere. A study by risk analysis group Verisk Maplecroft[4] said that eleven of the fifteen metropolitan areas most susceptible to sea level rise are in Asia. Guangzhou and Dongguan in China were worst placed, and others include Tokyo, Jakarta, Ho Chi Minh City and Shanghai. "Among the most prominent cities outside of Asia facing the highest risk are Dubai, Alexandria and New York," it stated.

So far, governments are most concerned with protecting their own coasts and citizens – with levees, dams, dykes and groynes (low barriers or walls built out into the sea to stem erosion) – rather than get into mind-boggling construction projects that would require unprecedented international cooperation.

Indeed, in many nations coastal protection is quietly evolving from hard walls to softer fixes. The Netherlands, where planes at the main airport at Schiphol touch down on runways that are 4 metres below sea level, is trying to enlist nature for help. In one experiment, the Netherlands built what is probably the world's biggest sandcastle on the coast at Ter Heijde, in the province of South Holland, in 2011.

The resulting Zandmotor,[5] called the 'Sand Motor' or 'Sand Engine' in English, totals 21.5 million cubic metres of sand and gravel – equivalent in volume to stacking about twenty Empire State buildings in New York on the beach – dredged up from the seabed about 10km offshore. Costing about €70 million, the project seeks to harness winds and ocean currents to spread the sand along the coast, building up beaches and creating sand dunes that will shield the region from rising seas. "The Sand Motor is a unique experiment, because it works *with* water instead of against it," the project's website claims. It was originally projected to help the coast for twenty to thirty years, short-circuiting an alternative of more frequent 'beach nourishment' – smaller amounts of sand when needed – or building dykes or walls.

A decade on, the official heading the project says it seems to be having a longer life expectancy. "It is still transporting sand, we now think it will last 30 to 40 years," Carola van Gelder-Maas, project manager of the Sand Motor for the Ministry of Infrastructure and Water Management in the Netherlands, tells me. On our video call during lockdown, she has a background photo showing sand dunes on the coast behind her. She said it is a common misconception that the Dutch preferred walls to keep out the sea. For the past three decades, she explained that official policy was consistently, "If you can use soft material like sand to protect your country, use the soft way".

"If it cannot be done, due to huge storms or big erosion, then you build revetments, or dykes, or levees, or dams, or piles. If it is possible to do it with sand, we do it with sand," she said. In recent years, grasses have started to grow on more of the Sand Motor, after a slow start, she said. Their roots help to anchor the sand and reinforce the beach, again trapping more sand and building up the dunes. She said she firmly backed using nature, rather than walls. Ideas such as building walls linking the continent to Britain were simply unnecessary when Plan A seemed to be working, she said, questioning a need for a "Plan B, C, D and E".

In addition, the Sand Motor has had big spin-offs. It has become an attraction for hikers, joggers, artists, kite-surfers, fossil hunters and bathers. The Sand Motor is now a hooked-shape peninsula with a shallow lagoon between the outer sandbank and the coast. I visited a few years ago and the sky was crisscrossed by red, yellow, green and orange kites as people battled a stiff breeze. A string of beach bars has sprung up along the back of the dunes. Fossil hunters have found everything from the tooth of a woolly rhino to bones of mammoths dating from 20,000 to 40,000 years ago. The fossils came with the dredged sand – from a region of the North Sea known as Doggerland that once was high and dry and connected Britain to continental Europe. Britain only became an island in about 6,000BC when ice sheets blanketing northern Europe during the Ice Age receded.

Jacqueline Heerema,[6] a self-styled 'conceptual artist and (sub) urban artist-creator', told me there had originally been too little involvement of artists, philosophers and ordinary citizens in the idea of adding an ephemeral part to a country on the coast. But in the past decade, many artists have been drawn to the Sand Motor – exhibitions hosted there have encompassed when artist Theun Karelse created bowls of 'fossil soup' including likely Stone Age foods such as reindeer meat and local plants.

"The Dutch are masters in disguising a cultural landscape as a natural one. We tend to design, construct, reconstruct nature to fit our needs," Heerema wrote in one study of the Sand Motor. She said it was a place where people could "stand on the shoulders of their ancestors" – those who had lived here when it was dry land. Heerema said that the Sand Motor was also an inspiration to focus on the risks of climate change and sea level rise in a nation that has historically relied on dykes built by the government to keep the water away. "In the mind of a lot of people the idea of climate change was something that we all delegated to the national government. In the past 15 years it has completely shifted. People are now more aware that it is also a personal challenge," she noted.

Overseeing the Dutch Sand Motor, van Gelder-Maas said other nations often expressed interest in copying the idea, including some parts of the US to protect coasts shown to be vulnerable after Superstorm Sandy in 2012. So far, the Sand Motor has one 'little sister', a 1.8 million cubic metre project[7] in Norfolk, east England, and roughly a tenth of the size of the Dutch project. Norfolk's cliffs have been crumbling into the sea for centuries as part of natural shifts, but storms have been vicious in recent years.

The British project was built in 2019 to protect natural gas terminals on a clifftop, as well as the nearby villages of Bacton and Walcott. Natural gas passing through the terminals, about a third of Britain's natural gas supplies, comes from offshore fields and from pipelines connected to the European mainland. The £20 million Norfolk sandscaping project is a buffer for the facilities, run by companies including Shell, Eni, Perenco and the National Grid, and is expected to last for fifteen to twenty years. They were spooked after a severe storm in 2013 gouged away up to 10 metres of the cliffs on which the gas terminals stand.

"It's a matter of buying time," said Jaap Flikweert, Flood and Coastal Management Advisor at Dutch engineering group Royal HaskoningDHV, which designed the project. The sand absorbs the force of the waves on the new beach before they reach the cliffs. "The communities of Bacton and Walcott were in a situation where there was potentially no time at all," Flikweert told me. "If there had been a big storm and the sea wall had collapsed the local authority would not have had the money to replace the sea wall … Those communities, the people, would have to leave."

"Broadly speaking, [the sand] is developing as we expected. We were always expecting losses over time," he explained. In winter, sand washes away from the coast, alarming some residents, but returns in the summertime. "That's just how beaches behave."

The project has created jobs, including for riders of jet skis equipped with sensors to monitor how the sand is shifting close to the beach in waters that are too shallow for survey ships. "These people have made their hobby into a job," he said. The local district council even has a section on its website[8] for frequently asked questions. These apparently include: "What is the Council doing to tackle the increase in dog fouling on the newly widened beaches?" (Answer: anyone spotting a pooping dog can report it on an online form.)

Like in the Netherlands, the project has helped the local economy with more tourists coming in before the pandemic lockdowns. Local house prices have risen, and some hotels are upgrading rooms and the local authority is building more car parks. Not everyone is happy. Some locals complain that the project used the 'wrong type of sand'.

"Our grievance isn't the beach. It's not the sea defences, it's using fine sand. The wind picks it up and blows it up the beach," said Mark Wright, who lives in a caravan by the beach in Bacton and likes to be known as 'Bell'. He has suffered repeatedly from sand clogging everything from his water heater to his motorbike. "I've had no hot water for months. I have an old transit van. The door locks have been knackered by sand. My bike was buried by sand. They recently moved 550 tonnes of sand from the pub car park. What impact the sand has had financially … is yet to be realised. I have to do stupid things like changing my letterbox. Its lock is full of sand. I've got two doors to the caravan. One of them I can't even move. The yard is covered in sand. You're walking in it, sleeping in it. It covers your clothes. The caravan's wrecked, the sand's broken the locks in my vans. I've a brand-new chainsaw and it's covered in sand. It's heart-breaking.

Absolutely bloody heart-breaking," he told me by phone. And he said the sand was a health risk, "I can't even go out there and breathe when it's blowing sand". In many ways he said he preferred the previous risks of storm surges, totalling three since the catastrophic winter storms of 2013. When a storm was forecast, he has driven a van inland and stayed in a safe place until the winds eased.

Unlike Bell, the Crown Estate, which manages the seabed from which the sand was dredged as well as large swathes of the British coast, praised the scheme. The sandscaping "has not only provided a high level of defence for the Bacton Gas terminal, but also provided a crucial lifeline to the villages of Bacton and Walcott, located downdrift of the project, giving them time to consider options for their long-term resilience against coastal erosion and climate change," it told me in a statement.

Flikweert said that the risks of windblown sand should ease as time passes. The lighter grains would blow away inland or out to sea and the bigger, heavier grains will then form a harder, more resilient beach. He said the idea of sandscaping often seemed less attractive to investors than building a wall, partly because building on sand always seems less durable than placing lumps of concrete on the coast. And sandscaping also requires complicated environmental studies to show that dredging up sand offshore will not be too harmful to marine life. "These are complicated, difficult decisions. It is never easy for a decisionmaker, or a funder, to do this. In general people are starting to see the wider benefits of working with natural processes," he said, pointing to welcome side-effects including tourism and recreation. Among other places in Britain, he said that Penzance in Cornwall, south-west England, was studying the idea of sandscaping.

Clive Schofield, head of research for the Global Ocean Institute at the World Maritime University in Sweden, praised the Sand Motor concept as an idea that could be applied more widely. "Rather than feed every beach, they are providing the natural system with supply, with ammunition, to help replenish and stabilise the coast which is a fantastic response." But, he said it would not work everywhere. It would be impractical to shore up low-lying atolls in the Pacific – the sand would probably just wash away too fast.

However, there is often political opposition to big engineering solutions for sea level rise, especially in nations such as the US where there is a political gulf,[9] with Democrats far more likely to support action to combat climate change than Republicans. Former Republican President Donald Trump, for instance, was dismissive after the *New York Times* reported that the US Army Corps of Engineers suggested options including a US$119 billion sea barrier to protect New York from storm surges.[10]

"A massive 200 Billion Dollar Sea Wall, built around New York to protect it from rare storms, is a costly, foolish & environmentally unfriendly idea that, when needed, probably won't work anyway," Trump tweeted, exaggerating the cost. "It will also look terrible, Sorry, you'll just have to get your mops and buckets ready." Mayor Bill de Blasio retorted, also in a tweet: "'Mops and Buckets' We lost 44 of our neighbors in Hurricane Sandy. You should know, you lived here at the time. Your climate denial isn't just dangerous to those you've sworn to protect – it's deadly."

Among the best-known coastal protection projects in the world is Venice's long-delayed barrier costing about €5.5 billion to protect the city from storm surges, prompted by the worst floods in the city's history in 1966. Also, London's Thames Barrier, which has protected the city since 1982, cost about half a billion pounds.

Elsewhere, many nations such as Singapore have reclaimed land or even built artificial islands – the Maldives, in the Indian Ocean, has used dredging to create the island of Hulhumale. China has controversially expanded islands such as Mischief Reef in the South China Sea – also claimed by Taiwan, the Philippines and Vietnam.

In the US, Florida is seeking ways to protect citizens from sea level rise in a low-lying region widely watched as a leader in American efforts to hold back the tides. By some estimates, about 85,000 people live less than 3ft above sea level in Miami-Dade county.[11]

And the region is also largely built on porous limestone, meaning that water can slowly seep in beneath any walls built to protect against storms. In a plan in 2021[12] to cope with rising seas in coming decades, Miami-Dade said the region had coped with a rise in sea levels of about 23cm since 1930 and that seas were expected to rise by another 60cm in forty years and will continue rising. "We are feeling the effects already," the report says.

Among ways to adapt, the plan aims to bring in compacted soil mined from other areas to raise the land on which new buildings, roads and other infrastructure are sited. It also advised putting buildings on stilts – a technique already used on buildings, including the Perez Art Museum in Miami – to enable flood waters from storm surges to pass underneath.

The plan also called for a shift of construction to naturally higher ground, for instance around the city's Metrorail lines. It encourages more waterfront parks to make room for water and recreation and to make room for more floodwater in yards, streets and parks.

The plan is a balancing act between action and not scaring away investors, including wealthy winter visitors from more northern states, often retirees, known as 'snowbirds'. It did not give a total cost for the proposals.

"It's very thoughtful – it has a lot of solutions that are going to serve them well in the next 10 to 20 years," Rob Moore, a sea level expert at the US Natural Resources Defense Council, said of the Miami plan. But in the longer term, he said that the US, the wealthiest nation on the planet, will have problems defending its cities if sea level rise accelerates. "Is the United States going to let lower Manhattan go underwater? Probably not. We'll do something there," he told me. "Are we going to do that in Savannah, Georgia, and Charleston, South Carolina, and Portland, Maine, and Norfolk, Virginia, and Tallahassee, Florida, and Brownsville, Texas, and Lake Charles, Louisiana? No, we're not. Even a country as rich as the United States doesn't have the ability to build a wall" along the entire coast.

Walls and other flood defences can widen racial and income inequity, he said, since they often favour affluent white people over black people and minorities. A conventional cost-benefit study would probably show it is worth building coastal defences to protect 100 beachfront homes worth US$1 million each – likely to be owned by wealthy white people – rather than to protect 100 coastal homes worth just US$100,000 each with a more racially diverse population, Moore said.

That theory has been backed up by research. Armouring the shoreline in North Carolina "correlates with higher home values, household incomes, and population density and low racial diversity," Harvard researchers A.R. Siders and Jesse Keenan found in one study in 2020.[13]

So far, the US has focused heavily on building up beaches by using sand dredged up offshore for 'beach nourishment'. Overall, it has spent almost US$11 billion on beach nourishment along its coasts totalling a billion cubic metres, according to research by Western Carolina University with data back to 1923.[14]

But current policies may not be enough, and there are hints that some investors are getting skittish about coastal property. One US study found that house prices in areas of coastal Florida most exposed to sea level rise declined from 2018–20 by about 5 per cent compared to those in safer zones, following a period of sluggish sales in waterfront areas since 2013. "We provide new evidence that the 'most liquid' parts of Florida (in that they are most likely to be underwater by 2100) have increasingly illiquid housing markets," wrote authors Benjamin J. Keys and Philip Mulder in a report for the National Bureau of Economic Research in 2020.[15]

Republicans are less likely than Democrats to be worried about global warming, and may be helping to support house prices simply because they doubt sea level rise will be an existential problem for low-lying areas on the coast.

Still, the Union of Concerned Scientists (UCS) says that ever more properties in the US are likely to be at risk as seas rise. "By the end of the century, homes and commercial properties currently worth more than US$1 trillion could be at risk," it stated in a study.[16] That same UCS report is based on extreme scenarios of sea level rise used by US government scientists – of a rise of about 60cm by 2045 and 2 metres by 2100. Other scenarios used by the US government are far less, down to just 30cm by 2100. The UCS said that people should be aware of the worst plausible threats to their homes from sea level rise, just as they insure against extremely rare but catastrophic fires.

Based on a worst case, the UCS said that more than "300,000 of today's coastal homes, with a collective market value of about US$117.5 billion today, are at risk of chronic inundation in 2045 – a timeframe that falls within the lifespan of a 30-year mortgage issued today". In that case, banks are likely to be increasingly reluctant to grant long-term mortgages to buyers in vulnerable areas. That could trigger default by property owners unable to repay what they have borrowed, especially if the value of their houses falls. In the end, taxpayers might have to prop up the mortgage bond market. Among other threats to the value of coastal property, local authorities around the world are likely to designate ever more coastal zones as flood zones – leading to stricter, and more expensive, building codes. Rising flood insurance policies could also be a spur for many people to move inland.

With a bewildering range of problems as seas rise, some of the world's leading climate researchers are starting to take engineering options more seriously.

Among the most drastic scientific proposals: pump vast amounts of sea water 700km inland onto Antarctica where it would freeze into ice and help stabilise sea levels. The new ice would slide back towards the ocean over centuries, but it would buy the world time to come up with long-term solutions to global warming.

Most people fail yet to grasp the seriousness of the threat of melting ice and rising seas, said climate scientist Levermann, who was one of the authors of the 2016 study[17] about coating Antarctica with more ice.

An extended period of 2°C of warming above pre-industrial times would mean, over time, 5 metres of sea level rise as Antarctica and Greenland thaw, Levermann said, based on records of what happened thousands of years ago when the world warmed between ice ages. Remember, the 2015 Paris Agreement aims to limit global warming to well below 2°C above pre-industrial times, while pursuing efforts for a 1.5°C limit. Temperatures are already up by about 1.2°C.[18]

Five metres of sea level rise would threaten cities around the world, and the cultural heritage built up over centuries. "Not keeping the Paris climate agreement really commits cities to the sea," Levermann told me. Unfortunately, he said, too many people reckoned that scientists exaggerate the risks.

He said it was clearly a terrible idea to mess up Antarctica, a home for penguins, whales and other wildlife and the only continent virtually untouched by humans. On the other hand, "I don't think that people who live in New York, or New Orleans, or Tokyo or Hamburg or London think that it's fine if in 200 years the city is gone," he said. "What we should ask is 'do we value Antarctica more than New York, Shanghai, Calcutta?' I think that's a really fair question," he proposes.

Levermann and colleagues estimated it would take about 1,000 years for water pumped 700km inland onto Antarctica to flow back to the ocean as ice in a glacier. "The approach offers a comprehensive protection for entire coastlines particularly including regions that cannot be protected by dykes," their study says. Among massive technical problems are energy supplies. Pumping enough water to offset sea level rise of 3mm a year up to a height of 4,000 metres – typical of inland Antarctica – would exceed 7 per cent of the world's energy supplies, they estimated. It would require the construction of hundreds of thousands of wind turbines to pump the water up in pipes. And the scientists' calculations of the energy needs omit some big heating demands, such as keeping the pipes warm to prevent the water from freezing before it reached its destination. Burst pipes several hundred kilometres inland in Antarctica would be a lot harder to fix than after a winter chill in a home in New Hampshire or Sweden.

Other environmental problems are that the plan would violate a 1991 environmental protocol[19] of the Antarctic Treaty, under which nations "commit themselves to the comprehensive protection of the Antarctic environment and dependent and associated ecosystems and hereby designate Antarctica as a natural reserve, devoted to peace and science".

Another option is to do a more targeted protection of the West Antarctic Ice Sheet, where ice may already have begun an irreversible melt into the Southern Ocean, Levermann said. Under that model, vast amounts of water could be pumped up onto the Thwaites and Pine Island glaciers where it would be released to freeze inland across an area about the size of Costa Rica. That extra ice could stabilise the most vulnerable ice, slowing its accelerating slide towards the sea.

The West Antarctic Ice Sheet region locks up ice equivalent to about 3.3 metres of sea level rise and is the most urgent focus of action since it is showing increasing signs of instability. By contrast, much bigger East Antarctica is generally much colder, and more stable.

Other scientists have drastic ideas[20] for geoengineering.

A group of scientists led by glaciologist John Moore at Beijing Normal University suggests engineering to slow glaciers in both Greenland and Antarctica. They say it is a trade-off of benefits and costs between glaciers and people. "Is allowing a 'pristine' glacier to waste away worth forcing one million people from their homes? Ten million? One hundred million? Should we spend vast sums to wall off all the world's coasts, or can we address the problem at its source?"

"Buttressing of glaciers needs a serious look," they wrote in a comment in the journal *Nature* in 2018. Among their proposals was to build a 100-metre high subsea ridge across a 300-metre deep fjord at the end of the fast-thawing Jakobshavn glacier in Greenland, where the ice is melting as warmer water from the Atlantic is seeping in and melting the glacier from beneath. The subsea wall on the seabed across the 5km wide fjord "would reduce the volume of warm water and slow the melting. More sea ice would form. Icebergs would lodge on the sill and prop up the glacier," they wrote. Ice spilling from the huge glacier accounted for about 4 per cent of sea level rise in the twentieth century. The underwater wall could be built by dredging up 0.1 cubic km of gravel and sand – a tenth of the amounts dredged up to build the Suez Canal – they reckoned. "It's not total science fiction," Moore said in a *Nature* podcast, referring to the scale of the project. He conceded that the initial reaction from many experts was "absolute horror". Another idea proposed by Moore's group is to build artificial islands around Antarctica, which might act to slow the flow of glaciers into the sea. A 300-metre-tall island built on Jenkins Ridge, a high point on the seabed by Pine Island glacier, could require 6 cubic kilometres of material, they estimated, but could slow the flow of the glacier. Risks include that such artificial islands would simply be bulldozed away by the ice sheet.

Broader geoengineering – deliberately trying to interfere in how the planet works – might also be used to keep the planet cool, despite many objections.

Among projects being mooted are balloons or planes that could spray tiny light-reflecting chemicals into the stratosphere, as a planetary sunshade. As briefly touched on previously, the idea is that it would help dim sunshine much as volcanic ash can dim sunlight – in 1991 an eruption of Mount Pinatubo in the Philippines threw up so much sulphur dioxide into the atmosphere that the gas caused global temperatures to fall by about 0.5°C until 1993. Such geoengineering has many unknowns – constantly spraying chemicals into the upper atmosphere would cause acid rain and could disrupt rainfall patterns, perhaps shifting monsoons vital to produce food.

But geoengineering may slowly be gaining support from some nations. The US National Academies of Sciences, Engineering and Medicine suggested[21] in 2021 that a five-year research project costing US$100 million to US$200 million to decide whether solar geoengineering was worth pursuing.

"This research program could either indicate that solar geoengineering should not be considered further, or conclude that it warrants additional effort," said Chris Field, Perry L. McCarty Director of the Stanford Woods Institute for the Environment, and chair of the committee that wrote the report.

So, governments have massive choices about how best to safeguard their coasts. One thing is obvious: deep cuts in emissions are best.

After the IPCC issues its bleak report about climate change in 2021, I get back in touch with Groeskamp about his suggestion of vast dams to protect Europe.

He now feels that the project would become interesting for coastal states with major low-lying cities by the shore – like Amsterdam in the Netherlands, Copenhagen in Denmark, or London in England – if 2 or 3 metres of sea level rise looks inevitable.

If the IPCC worst scenarios turn out to be correct, "we should definitely start looking at such solutions," he said. "This type of thinking is becoming less out of the box as we go along ... I'm finding that some people don't think it's a crazy idea at all."

10

"CLIMATE REFUGEES?" A PACIFIC QUEST MAY OPEN THE WAY

I'm the same as people who are fleeing war.
Those who are afraid of dying, it's the same as me.

Ioane Teitiota, a Pacific islander from Kiribati who lost an appeal for asylum in New Zealand after he was threatened by rising sea levels

Sea level rise may force millions of people to flee to higher ground in coming decades, but migrants lack clear rights and protection under the UN's 1951 Refugee Convention. Slowly, things are changing – a UN committee opened the door to climate change asylum claims in 2020. Low-lying island states are also working to ensure they retain fishing rights, even statehood, if they are swamped by rising waters.

A DECADE AGO, A TRAFFIC patrol officer in New Zealand pulled over a driver because of a faulty rear light.

The stop was the improbable start of a string of lawsuits that reached the New Zealand Supreme Court and triggered a landmark ruling by the UN Human Rights Committee[1] in 2020 expanding the safety net for people fleeing abroad because of climate change. The fallout of the broken light may make it easier in future for people whose lives are threatened by global warming to seek asylum abroad.

The driver pulled over was Ioane Teitiota, a citizen of Kiribati, the nation of thirty-three islands scattered across the central Pacific Ocean halfway between Australia and Hawaii, who was living in New Zealand.

Teitiota had come in 2007 to New Zealand and got a job as a vegetable farm worker while his wife, Angua Erika, worked as waitress. Together they had three children in New Zealand, who were not New Zealanders after the nation in 2006[2] scrapped an automatic right for anyone born in the country to be a citizen.

Teitiota was detained after the traffic stop under an arrest warrant issued because he had failed to renew a permit that allowed the family to stay in New Zealand until October 2010. That meant he was illegally in the country. Ineligible for a new visa after missing the deadline for renewal, Teitiota sought asylum, saying that Kiribati was facing steadily rising sea levels because of climate change that could force the inhabitants to abandon the islands in coming decades.

What became a high-profile quest over several years to become the world's first 'climate refugee' ultimately failed in the New Zealand Supreme Court.[3] It was the first such case to reach a top national court anywhere in the world at a time when governments in rich nations are reluctant to widen the scope of refugees to include millions of people suffering the impacts of global warming, from the Middle East to Central America.

Teitiota, born in the 1970s, and his family were deported back to Kiribati in 2015, where he now lives in the capital on Tarawa island. About 120,000 people live in Kiribati and its low-lying coral atolls are among the most vulnerable in the world, with an average elevation of less than 2 metres[4] above sea level.

Teitiota and his New Zealand lawyer Michael Kidd then appealed to the UN Human Rights Committee against his deportation. In 2020, the committee sided with New Zealand in ruling that Teitiota's life was not under threat from climate change. At the same time, it also spelt out that "countries may not deport individuals who face climate change-induced conditions that violate the right to life".

"Historic UN Human Rights case opens door to climate change asylum claims," the UN's Office of the High Commissioner for Human Rights declared in a statement about the ruling.[5] Filippo Grandi, the head of the UN refugee agency UNHCR, said the ruling meant that "if you have an immediate threat to your life due to climate change, due to the climate emergency, and if you cross the border and go to another country, you should not be sent back because you would be at risk of your life, just like in a war or in a situation of persecution".[6] "We must be prepared for a large surge of people moving against their will," he said in 2020. "I wouldn't venture to talk about specific numbers, it's too speculative, but certainly we're talking about millions here."

For Teitiota, of course, that is no help for him in Kiribati, on Tarawa island and sometimes on the remote island where he was born, which is about three days away from Tarawa by boat.

Teitiota had argued in vain to the UN Committee that his family's right to life was at risk in Kiribati. One of his children had suffered from "blood poisoning, which caused boils all over his body", salt water made it hard to grow crops, drinking water was scarce and there was a risk of drowning in storm surges. Teitiota says his family is a lot worse off. "Forgive my ignorance, but to be frank, I'm quite disappointed with the outcome of my case" in the UN, he told the Australian Broadcasting Corp[7] after the 2020 ruling. "The notable difference is that my children are more vulnerable here to the spread of diseases, such as viral infections like the flu, or diarrhoea," he said. "What's important to me is for big countries to consider the challenges we face with respect to climate change ... Personally I think big countries like New Zealand should accept us and not ignore our plight because our islands are very low lying and we are vulnerable to the slightest bad weather or storm surge so our life here is being affected in one way or another. So, I want to ask these big countries to please take our case seriously because we need their help."

His lawyer had argued to the UN Committee that "the situation in Tarawa has become increasingly unstable and precarious due to sea level rise caused by global warming. Fresh water has become scarce because of saltwater contamination and overcrowding on Tarawa. Attempts to combat sea level rise have largely been ineffective. Inhabitable land on Tarawa has eroded, resulting in a housing crisis and land disputes that have caused numerous fatalities. Kiribati has thus become an untenable and violent environment for the author and his family."

I tried to get in touch with Teitiota, via local environmental activists and journalists, to see how he and his family are getting on after the UN ruling, and if they have any other legal avenues, but without luck. He has generally kept a low profile – some in Kiribati accuse him of smearing the nation's reputation around the world by making it sound as if Kiribati is riven by strife and all but uninhabitable.

Of the eighteen-member UN Committee, two dissented and said New Zealand should have allowed Teitiota and his family to stay. One of the two, Duncan Laki Muhumuza, said the committee was wrong to argue that many people in Kiribati were already coping with the risks that Teitiota complained about. "The action taken by New Zealand is more like forcing a drowning person back into a sinking vessel, with the 'justification' that after all, there are other passengers on board. Even as Kiribati does what it takes to address the conditions, for as long as they remain dire, the life and dignity of persons remains at risk," he wrote.

Even so, Matthew Scott, of the Raoul Wallenberg Institute of Human Rights and Humanitarian Law in Lund, Sweden,[8] said the committee's decision broke new ground partly by insisting that individuals have a "right to life with dignity", broader than a mere "right to life" laid down in international treaties. That could mean a lower bar to become a refugee. "This idea of the right to life with dignity is relatively new," he told me by phone. Refugees no longer have to establish that they risk death if they are sent home but that they "will not have adequate shelter, access to water, health". He also said it was a landmark because it was the first time that the UN Committee had been so explicit about linking climate change and asylum.

The New Zealand Supreme Court had made a similar point. It said that the lower courts' repeated denials of asylum to Teitiota "should not be taken as ruling out" the possibility that climate change or other natural disasters could pave the way to refugee status for others in the future.

Law professor McAdam at the University of New South Wales, said the Human Rights Committee had pushed the door ajar for asylum claims. The ruling meant both that "as climate impacts worsen, future similar claims might well succeed" and that "potentially even now, a different person somewhere in the world might already have a valid protection claim," she told an online seminar in 2021.[9]

Teitiota's pleas fell short partly because the UN Committee said that there was still some time left to solve his plight. Teitiota argued that Kiribati might become uninhabitable within ten to fifteen years as seas rise. The committee said that meant there was still an opportunity for Kiribati, with international help, to "take affirmative measures to protect and, where necessary, relocate its population".

As early as 1990, the IPCC warned that "the greatest single impact of climate change could be on human migration".

Andrew Harper, special advisor on climate change to the UN refugee agency, UNHCR in Geneva, said most people displaced by natural disasters prefer to stay in their own countries rather than cross an international border where they would be considered refugees. UNHCR is working to bolster regional refugee agreements, including by nations in the Pacific, Latin America or Africa, rather than negotiate new, global accords to cover the impacts of global warming on migration, he said. "Let's not try to bring in another bureaucratic layer from New York or Geneva to complicate things," he told me. "It's better to work with communities on the ground and see what they want."

He said the UNHCR was also doing more to study climate change, from slow-moving sea level rise to storms, droughts, floods and heatwaves, to help predict when people will be forced to move. "There is a lot of information about glaciers and temperatures but it's not often brought to the next area – what will that do to communities, what will it do to conflict, violence, displacement?" he said, adding that 90 per cent of refugees were from nations on the frontlines of climate change with the least capacity to adapt.

"I have never been so close in looking at scientific literature in trying to understand climatic conditions," he said, mentioning the quickening pace of melting ice in Greenland and Antarctica. "If we take science into account we should never be in a situation where we say, 'oh wow, we didn't expect that'" if there is a sudden surge of migration. "Climate is evolving around two elements: you are going to have too much water in certain places, or you're going to have too little," he said. In Somalia, for instance, UNHCR tracks the price of drums of water – a price rise could reflect a drought, or perhaps stockpiling by armed militias, and presage people fleeing to neighbouring Ethiopia or Kenya.

Worldwide, in the past decade, "weather-related events triggered an estimated 23.1 million displacements of people on average each year," the WMO, the UN weather agency, said in a 2021 report.[10] In 2020 alone, for instance, 2.4 million people were displaced in India and 2.5 million in Bangladesh, after Cyclone Amphan struck the region straddling the two countries. Damage to homes "likely resulted in homelessness and prolonged displacement for many thousands," the WMO said.

The number of people in need of humanitarian aid because of climate-related disasters – storms, droughts and floods – could double to more than 200 million people a year by 2050 from 108 million today, a report[11] by the International Federation of Red Cross and Red Crescent Societies (IFRC) showed.

Many seeking asylum already argue that climate change is at least a factor in forcing them to abandon their homes. People fleeing Central America to the US have sometimes blamed drought or the impact of hurricanes. Some researchers[12] have linked the civil war in Syria, which sparked a refugee exodus since 2011, to long-term drought and man-made climate change. But plaintiffs like Teitiota have yet to persuade courts that climate change is powerful enough as a core argument to secure asylum.

The UNHCR, and vulnerable nations, do not use the term 'climate refugees'. Small island states, for instance, say that calling people like Teitiota 'climate refugees' is patronising because it wrongly puts the stigma on the victim, rather than on the industrial nations that are spewing out greenhouse gas emissions and causing harm.

UNHCR says, "It is more accurate to refer to 'persons displaced in the context of disasters and climate change' than to 'climate refugees'."[13] The IPCC also argues it is hard to identify climate migrants. "Environmental conditions and altered ecosystem services are few among the many reasons why people migrate. So while climate change impacts will play a role in these decisions in the future, given the complex motivations for all migration decisions, it is difficult to categorize any individual as a climate migrant," it says.[14]

And using the word 'refugees' for people uprooted by climate change could turn the UN's 1951 Refugee Convention[15] on its head. The convention defines a refugee as "someone who is unable or unwilling to return to their country of origin owing to a well-founded fear of being persecuted for reasons of race, religion, nationality, membership of a particular social group, or political opinion". Climate change isn't mentioned, partly because the idea that greenhouse gas emissions could cause crop failures, rising sea levels, floods and water shortages was unknown in the early 1950s. Plus, unlike traditional refugees fleeing persecution in their home nations, people like Teitiota want to move to a nation that is the source of the greenhouse gases that are 'persecuting' them. In court, Teitiota's lawyer argued that rich nations such as New Zealand were responsible for a generalised persecution by emitting greenhouse gases.

The New Zealand courts flatly rejected the argument. "This completely reverses the traditional refugee paradigm," Justice John Priestley wrote in the New Zealand High Court, dismissing Teitiota's claims. "The claimant is seeking refuge within the very countries that are allegedly 'persecuting' him." Accepting the argument that greenhouse gas emissions are a form of persecution would open the floodgates to "millions of people", Priestley argued.[16]

Scholars believe it would be difficult, if not impossible, to renegotiate the Refugee Convention formally to add 'climate change' as a form of persecution. Scott at the Raoul Wallenberg Institute, for instance, said, "Rewriting the refugee convention is a bad idea. The general perspective is that it is very likely it would result in a watering down of protection rather than an improvement."

Still, the world is badly prepared for a worsening migration crisis as climate change disrupts food and water supplies. Many nations have been hostile to immigration this century – former US President Donald Trump won the 2016 election with a signature promise to get Mexico to build a wall along the US southern border to keep immigrants out. President Joe Biden plans a more humane approach.

Beyond the puzzle of how the world will manage future refugees, rising sea levels also raise questions about 'what is a state?' If a country disappears beneath the waves, can it still exist, like some sort of Atlantis?

Small island states have been warning that they risk existential threats for decades, ever more stridently since negotiations on climate change began in the 1990s. 2021 will see the twenty-sixth of these Conferences of the Parties or COPs – COP1 was in Berlin in 1995. Over the years, I've been to thirteen of these marathon events.

At one of the first I attended, COP12 in Nairobi in 2006, Pacific island delegates warned that sea level rise might require "taking down national flags outside UN headquarters in New York". In 2009, the President of the low-lying Maldives in the Indian Ocean highlighted the risks by donning scuba gear and holding a cabinet meeting under water,[17] at submerged desks on the sandy seabed, to send an appeal for action to a climate summit in Copenhagen later that year.

The most immediate issue is how to safeguard maritime rights that are key to fisheries and mining offshore as oceans rise. Many small island nations prefer to be known as 'Large Ocean States' to show how the existence of remote, low-lying islands ensures rights to large swathes of the ocean.

All coastal states have rights to an 'Exclusive Economic Zone' (EEZ) 200 nautical miles (370km) from their shores, guaranteeing access to fisheries or offshore mining, under the 1982 UN Convention on the Law of the Sea (UNCLOS).[18]

Imagine a small inhabited island of 1 sq km in a remote part of the Pacific – an EEZ stretching in all directions gives it control over an ocean area of about 430,000 sq km, bigger than the land area of Germany or the US state of California. That is valuable, largely because the Pacific has the world's richest tuna stocks, worth billions of dollars a year.

As seas rise, a big risk is that islands may no longer exist and lose their EEZs. Most immediately, Article 121 of UNCLOS says, "Rocks which cannot sustain human habitation or economic life of their own" do not qualify for an EEZ. People might leave remote islands as rising seas cause ever more floods, swamping homes and washing salt water onto crops, long before they sink beneath the waves. That could demote 'islands' to 'rocks' under UNCLOS' definition. Foreign fishing fleets from nations such as South Korea, Japan and the US might be able to turn up, catch tuna off a recently abandoned island and argue that what was once an EEZ is now part of the high seas around a 'rock', open to all. A Pacific government might try to hit back by arresting foreign fishing vessels, risking an international dispute.

Pacific islands are acutely aware of the risks. "Peace and security are threatened by climate change. It's an existential threat. It's not just about the islands disappearing, it's about peace and security," said Jens Kruger, Deputy Director of the Ocean and Maritime Program at the Pacific Community,[19] based in Suva, Fiji, which represents Pacific governments.

In the real world, ten Pacific nations including Fiji, Vanuatu, Marshall Islands and Tuvalu, have a population of about 2 million on a land area of 62,000 sq km, roughly the size of Sri Lanka or the US state of Florida. By contrast, their offshore rights total a gigantic 40.6 million sq km for fishing and mining – bigger than the surface of the Moon at 38 million sq km or the combined land areas of Russia, Canada and China, according to the UN Development Programme.[20]

So Pacific nations are quietly getting ahead of the problem by mapping existing islands and locking in their EEZs as final, irrespective of future sea level rise, to head off potential conflicts over fishing rights. The idea is that the EEZs will remain, even if the islands that generate them disappear. "The fact that the sea level is rising is not really our doing. We are disproportionately affected by it. But we are not bystanders, waiting for other countries," Kruger said. Pacific governments are mapping the outer limits of their 200 nautical miles (nm) zones around remote islands and submitting lists of the precise geographical coordinates of these zones to the UN. Kruger said Kiribati, Marshall Islands, Niue, Samoa, Tuvalu, Australia and others either have changed, or are in the process of changing, their national legislation to allow them to use geographical coordinates to stake out their EEZs. "We have thousands of islands. We are committed to UNCLOS and use the low-lying atolls to generate our maritime zones, and once these are delineated in accordance with UNCLOS they cannot be challenged or reduced as a result of sea-level rise and climate change," he told me.

The changes mean that EEZs will become dislocated from the reality of islands shown on maps as seas rise, which could be a problem for shipping in future. Kruger said, however, that sandy atolls shift naturally so ships already have to be wary of trusting charts. "We have many low-lying islands and some of these could be really remote – 2–3 days by ship away from the capital. But nobody really knows what is happening," he said. "Sometimes a cyclone erodes an island away, the currents might be so strong that the sand is shifted, the salt spray is so strong that it kills off the plants. Sand bars can be re-established after extreme events and plants can grow back. It's a dynamic system, it changes over time," he said. The newly mapped "maritime zones, however, are considered permanent".

Some climate change sceptics or deniers often point to such natural processes – some islands naturally sink while others rise up – to argue that global warming may be a hoax or exaggerated. But the island nations' drive to lock in EEZs is compelling evidence that the nations most vulnerable to climate change view sea level rise as an existential threat. There would be little incentive to make the massive effort to lock in permanent maritime zones if they reckoned EEZs will be stable this century.

For the low-lying Pacific island states, the right to sell licences to foreign fleets to catch tuna in EEZs is a vital source of revenue. The Western and Central Pacific Fisheries Commission[21] estimated the region's tuna catch of 2.9 million tonnes in 2019 was worth US$5.8 billion and accounted for 55 per cent of the global total.

But other nations have their eyes on tuna stocks and EEZs. Former Australian Prime Minister Kevin Rudd once suggested[22] that low-lying Pacific island nations such as Tuvalu, Kiribati and Nauru might want to trade their EEZs for citizenship abroad.

"If our neighbours requested this, and their peoples agreed, Australia would become responsible for their territorial seas, their vast Exclusive Economic Zones, including the preservation of their precious fisheries reserves," he wrote on his website in 2019, several years after leaving office. "Under this arrangement, Australia would also become responsible for the relocation over time of the exposed populations of these countries … to Australia where they would enjoy the full rights of Australian citizens." He added, "If these countries start to submerge in the years ahead, Australia would face international pressure to provide safe haven for our Pacific neighbours anyway."

Tuvalu's then Prime Minister Sopoaga[23] angrily shot down the idea. "The days of that type of imperial thinking are over," he told the Australian Broadcasting Corp. "Certainly [we will] not to be subjugated under some sort of colonial mentor; those days are over. We are a fully independent country, and there is no way I'm going to compromise our rights to fisheries resources, our rights to our immediate resources," he said.

But some commentators said the small island states should be less dismissive.

"You can't eat sovereignty, you can't drink independence, and you can't build a house on a flag floating in the middle of the ocean," wrote journalist Bruce Hill, in an opinion article[24] published by the Lowy Institute, a leading Australian think-tank. "If this offer was an opening bid in a renegotiation of the status of the citizens of Nauru, Tuvalu, and Kiribati in the light of expected climate change, they might like to consider it a little more seriously than Sopoaga has," he added. "It may sound cynical, but as time goes on, their bargaining position will probably only get weaker as sea levels rise around them," he wrote.

By contrast, many legal experts firmly back the small island developing states.

Schofield, the head of research for the Global Ocean Institute at the World Maritime University in Sweden, said that there was a widening support for the approach taken by Pacific states to safeguard their EEZs as a linchpin for their economies. "There is an equity and justice issue to weave into this," he told me, saying that people living in Pacific island states were among the smallest contributors to global warming on the planet and needed international support. Unlike foreign fleets, Pacific islands lack the technology to freeze and transport tuna to Tokyo, for example, where sushi commands sky-high prices. "But they do have the conservation burden – the rights over the resources but also the burden in terms of the responsibility to preserve the marine environment and to safeguard the health of the stock," Schofield said. He said it was unclear what would happen if a foreign fleet were to start fishing off a remote newly uninhabited atoll and claim it no longer qualified as an island. In international law "that is, as yet, untested", he said.

Among other uncertainties, is the question of whether the growth of corals, which underpin islands, can keep up with the rise of sea levels. Globally, annual sea level rise has been about 3.7mm in recent years. "In a healthy ocean, coral can grow by up to 11mm a year. That far outstrips the present rate of sea-level rise. But that rate is in a healthy ocean before the pre-industrial age. We have substantial uncertainties about whether coral can keep up in a warmer ocean, a more acidic ocean and a de-oxygenated ocean," Schofield said.

David Freestone, a law professor at George Washington University who wrote a pioneering report in 1991 about 'International Law and Sea Level Rise', said the issues around rising oceans had since "become much more pressing" with ever more meltwater from Greenland and Antarctica. He noted that UNCLOS does not spell out whether states have an obligation to adjust maritime boundaries if islands disappear. And he told me on a Zoom call that he knew of no examples of nations that had lost an island – perhaps because of an earthquake, or erosion – and with it an EEZ.

The International Law Association (ILA),[25] grouping thousands of experts who work to clarify and develop laws, passed a resolution in 2018[26] siding with the Pacific states' drive to fix their maritime zones as seas rise. The views of the ILA are often used by governments as a guide. As long as measurements were made in line with UN rules, it said that "baselines and limits should not be required to be recalculated should sea level change affect the geographical reality of the coastlines".

The Pacific island state of Tuvalu has gone one step further, saying that any state which has diplomatic relations with it automatically recognises Tuvalu's self-declared EEZs. "We need not wait for the global community to respond to our calls; we can take action now to secure the limits and future of the Blue Pacific," Simon Kofe, Foreign Minister of Tuvalu, said in a 2020 speech[27] outlining that condition.

While most of the focus is on sinking islands, there are also disputes about whether countries can build up rocks into artificial islands to gain new rights to fishing and offshore mining or oil and gas exploration. In a landmark case, an international tribunal in The Hague in 2016 criticised[28] China's construction of artificial islands in the South China Sea and said they could not be used to claim sovereignty in the region.

The Philippines brought the case after China built up rocky outcrops including Mischief Reef, which is also claimed by Manila, by dredging up sediments. Beijing argued that the outcrops are 'islands', not mere 'rocks' in international law, and thereby qualify for maritime zones for fishing and mining.

The tribunal dismissed China's views. "A rock cannot be transformed into a fully entitled island through land reclamation," it concluded. While the case only applies to the Philippines and China – which denounced the ruling – it could have wider implications in future legal disputes.[29]

Details of the ruling might, in the long term, help states at risk of rising seas, Schofield and Freestone wrote in a joint article.[30] They noted, for instance, that the tribunal made it clear that defining a rock or an island had to be "assessed on a case-by-case basis". And the tribunal also made it clear that to qualify as an inhabited island – able to claim an EEZ – the population "need not necessarily be large" and "in remote atolls a few individuals or family groups could well suffice". It also found that an island did not need to be inhabited permanently to qualify for an EEZ, saying that island-hopping nomadic peoples "could also constitute habitation".

Further down the line is the even more fraught legal issue of whether states could disappear entirely as seas rise. Legal scholars say that the 1933 Montevideo Convention,[31] which defines statehood, may have some wiggle room that could be used to help safeguard low-lying nations as sea levels climb.

Article One of the convention lays out the bare-bones requirement for statehood:

> The state as a person of international law should possess the following qualifications:
> a) a permanent population
> b) a defined territory
> c) government; and
> d) capacity to enter into relations with the other states.

But it doesn't, for instance, define a minimum threshold for population or territory.

Freestone, at George Washington University, noted that there are plenty of places that don't meet all four of those conditions. The Vatican City, for instance, run by Pope Francis as head of the Roman Catholic Church, is recognised as a state with observer status at the UN. But it has a population of only 800 and a miniscule territory in the heart of Rome of 0.44 sq km. It could easily fit inside Regent's Park in London or New York's Central Park.

"The Vatican has an established territory, a very small one, in the middle of Rome, but it doesn't really have a 'permanent population'," Freestone said, noting that the population included many Catholic men sworn to celibacy. "It has a government and can enter into international relations."

Freestone, one of the leaders of the International Law Association Committee on International Law and Sea Level Rise, added that there was a discussion among the experts about whether a requirement for a 'defined territory' necessarily meant 'land' or might be extended out to 'sea'. If maritime zones count as 'defined territory', a government might be able keep existing even if its land is submerged, Freestone explained. Such a solution would also enable it to retain fishing or mining rights offshore.

Davor Vidas, a professor at the Fridtjof Nansen Institute in Oslo who chairs the ILA sea level rise committee, also said that sea level rise could upend definitions of statehood if low-lying nations disappear and their people are forced to migrate. He said that having a population – even in exile – might be a more important key to statehood than territory. Some governments-in-exile, such as the Cambodian government in the 1980s, have been allowed by the UN to sign treaties.

"Ultimately the key factor may be the population better than the territory. Small island states are likely to become uninhabitable long before they become physically submerged," he told a video conference in 2021.[32] "These questions have far-reaching implications."

So, legal experts say that low-lying states might continue to exist by stretching the definitions of statehood under the Montevideo Convention even if they are swamped, perhaps on an artificial island, or as an enclave in another sovereign state. "It is possible to see that the Montevideo criteria are not merely a stringent set of rules that must be fulfilled, but rather that there is some fluidity and leeway," according to a study of statehood[33] by Rosemary Rayfuse of the University of New South Wales and Emily Crawford of Sydney Law School.

And yet the situation for small islands differs from many previous states that have broken apart. Countries formed from others, such as Serbia or Croatia established after the break-up of former Yugoslavia, were on the same land. Serbs and Croats were not forced to pick another part of Europe on which to establish a new state.

However, Rayfuse and Crawford noted that it is possible to have some of the characteristics of a state, even without territory. For instance, the Sovereign Order of Malta,[34] a lay religious order of the Catholic Church founded in Jerusalem in the eleventh century, has diplomatic relations with more than 100 states and permanent observer status at the UN. The order, now based in Rome, used to have territory in Malta but was forced out by Napoleon Bonaparte in 1798. "Given the dynamism and flexibility demonstrated by the international community and international law, it is reasonable to hypothesise a system whereby states that 'lose' key indicia of statehood, through no fault of their own, continue to retain the benefits and privileges of statehood," they wrote.

Overall, scholars reckon that it would be hard, if not impossible, to renegotiate UN treaties to overhaul issues of sovereignty, the Law of the Sea, or the treatment of refugees. All were written in eras before climate change became what the United Nations calls an existential threat to humanity.

But at least scholars see hope of interpreting these outdated texts in ways that will help the world adapt, especially the most vulnerable, low-lying nations. And it is in the interests of all coastal states, rich or poor, to safeguard their fishing zones if rising seas erode the land that is the baseline for offshore claims.

Sadly, none of that is any help to Teitiota who, but for a fateful traffic stop, might still be living in New Zealand. He told the BBC[35] in 2015: "I'm the same as people who are fleeing war. Those who are afraid of dying, it's the same as me."

EPILOGUE: 28 TRILLION TONNES OF ICE

IMAGINE VISITING A FJORD IN Norway or New Zealand where jagged snow-capped mountains surge from the sea. Go down to the shore and fill an 11kg bucket with water. Then hike up a steep path to the glacier above, pour the water out and leave it freeze into a lump of ice to make a symbolic, personal contribution to slow sea level rise.

There's a catch. In the real world you would have to extract and freeze a bucketful of water from the ocean once an hour, every day, year round to offset the average person's share of the current great melt.

Eternally clambering mountains with buckets of water is of course impossible and absurd, a Sisyphean gesture against the quickening thaw. But maybe the fantasy buckets of water give a glimpse into the mammoth scale of the task. After all, former US President Donald Trump once famously dismissed proposals for a sea wall to protect New York from sea level rise by telling people to "Get your mops and buckets ready!"

Writing this book, I've often felt that the melt of ice is too colossal, abstract and far away to attract the public attention it deserves as the planet heats up. The thaw simply doesn't translate into a sense of personal responsibility for causing the problem. And for almost all of the world's population the great melt is out of sight in remote polar regions in Greenland or Antarctica, or up in high mountain glaciers in the Alps or the Himalayas. Few people, like those in Iceland or the Peruvian Andes, live close to glaciers where the melt stirs emotions ranging from sadness to foreboding.

I've tried to focus on stories told by people on the frontlines – on low-lying islands in the Pacific or in the Caribbean or along the coast of Florida, the Netherlands or Britain. They are already acutely aware of sea level rise, and how each millimetre counts. Thankfully, young people, who risk living with a legacy of a metre of sea level rise by 2100, are becoming far more aware of the challenges.

Showing the scale of the problem, a 2021 study estimated that the world lost what scientists called "a staggering 28 trillion tonnes of ice between 1994 and 2017".[1] Of that total, their data shows that glaciers and ice sheets on land, which cause sea level rise as they melt into the oceans, lost 779 billion tonnes of ice on average every year in the 2010s.

With almost 8 billion people on Earth today, 779 billion tonnes works out at about 100 tonnes for each person, every year in the 2010s. In turn, that represents about 270kg per person every single day, or just over 11kg every hour. If that amount of water was leaking from a pipe into your house, you'd get it fixed immediately. Some of the historical melt has been caused by long-term natural trends, but our greenhouse gas emissions from fossil fuels are now the overwhelming driver of the warming planet.

Andrew Shepherd, professor of Earth Observation at the University of Leeds who was one of the authors of the study about the 28 trillion tonnes of ice, said the team produced what they thought was a shocking illustration of a trillion-tonne ice cube to ram home the message at public presentations.[2]

It shows a vast cube with sides 10km long by New York, smothering part of New Jersey and casting a long shadow over Manhattan, dwarfing its skyscrapers. To his surprise, "Some people thought it didn't look that big at all," he told me of reactions. "There was a feeling that if it can't be seen from space, then it's not impressive."

"Since then I have asked 'what does it mean for each individual?' Then it does hit home," he said. "The idea of having to repair the damage of ice loss by taking water out of the sea is more relatable," he said.

He now sometimes illustrates the task using the video game Minecraft, where water and ice come in blocks of a cubic metre, or 1 tonne. "It can be quite impactful to see that each person on Earth would need to shift 100 of these blocks from the ocean and back onto glaciers each year," he reflects.

And Shepherd's team says the 779 billion tonnes lost from land ice are only part of the problem of the great melt. Another 480 billion tonnes of ice are also vanishing every year as sea ice on the Arctic Ocean and around Antarctica melts away and ice shelves crack up. This ice is already floating on the sea and so does not contribute much to sea levels when it thaws. Taken together, the Earth lost 1.3 trillion tonnes of ice a year in the 2010s, quicker than in previous decades in the study. Separately, the oceans are swelling because water expands as it heats up.

The fantasy of everyone extracting water from the ocean with buckets – 11kg an hour per person – overlooks the massive differences in responsibility for causing climate change. Around the world, each person emits the equivalent of 6.8 tonnes of carbon dioxide a year, according to the US-based World Resources Institute think-tank.[3]

Americans emit 18 tonnes per capita every year, almost triple the world average and the most of any major nation. In a fair daydream, they would all be hauling almost three buckets of water out of the ocean every hour. Even if the US halves its emissions, Americans would still bear more responsibility than most people on Earth.

In this imaginary world, everyone in China (8.5 tonnes of carbon dioxide each) and the European Union (7.1 tonnes) would be carrying a bucket more than once an hour. Meanwhile, people in India, who emit just 2.5 tonnes of greenhouse gases a year, would be responsible for a bucket less than every two hours while citizens of the poorest nations in Africa with the lowest emissions would be able to relax, relatively.

Plus, those numbers of buckets do not reflect historical responsibility – industrialised Western nations have been burning fossil fuels for longest and so should be doing far more to keep their promises of leading the way to a sustainable future that will eliminate poverty and hunger and ensure equality for all people.

The risks of sea level rise are increasing, according to almost everything I have read while working on this book – there are 'doomsday' glaciers in West Antarctica, 'time bombs' locked in ice sheets, perhaps an 'irreversible' melt. Rare, severe coastal floods caused by storms may become annual events in many people's lifetimes.

It is hard to know how to react to the IPCC's alarming findings in 2021, such as that sea level rise exceeding 15 metres cannot be ruled out by 2300 with high emissions. We can't "rule out" a lot of things in life, like the chance that I'll be struck by lightning this week. Usually, we don't even bother to worry about such remote possibilities.

Still, when UN Secretary-General Guterres refers to the IPCC report as "code red" for humanity, it seems time for bolder action to cut emissions and to fund more science to understand the risks of a disintegration of Antarctica's ice. It is all about the most vulnerable people on the frontlines.

In assessing remote possibilities in the climate system, governments might remember former U.S. Vice President Dick Cheney, who formulated what became known as the "One Percent Doctrine" for dealing with terrorism. "If there's a one percent chance that Pakistani scientists are helping al-Qaeda build or develop a nuclear weapon, we have to treat it as a certainty in terms of our response. It's not about our analysis ... It's about our response," he said.

But doubts about the human cause of global warming are likely to linger long after governments, even OPEC nations, agreed the IPCC conclusion that the link is "unequivocal".

Santer, the scientist who helped identify the human fingerprint of climate change in a landmark 1995 report, said that "climate denialism is still alive and well in the United States Congress".

"But the times are changing," he told me after the IPCC report. "The concerning changes in extreme events - particularly heat waves, drought, floods, and wildfires – are diminishing the space in which denialism can thrive. It's tough to deny the reality of climate change when much of your country is on fire; when people are suffering and dying from record-shattering heat."

In 1990, around the time that governments were waking up to the risks of man-made climate change, the Stockholm Environment Institute suggested[4] three ways of gauging global warming: sea level rise, temperature rise and the amount of carbon dioxide in the atmosphere. "Such indicators could provide decisive evidence that climate is changing and could provide a yardstick for measuring progress in reducing the greenhouse effect," it said.

For sea level rise, the scientists proposed "a maximum rise of between 20mm and 50mm per decade". The lower bound has already been surpassed. Of the three yardsticks, sea level rise never caught on, partly because of the gigantic uncertainties in predicting how the ice sheets in Antarctica and Greenland will react to rising temperatures. The 2015 Paris Agreement, for instance, set goals only for limiting temperature rises.

In many ways, it's a shame that melting ice and sea level rise were omitted from international goals, because the thawing and freezing of ice as temperatures pass 0°C is a very visible sign of whether the world is warming or cooling. "Ice was the first thing to show the impact of climate warming," Shepherd notes. "If you hope to see a better trajectory, then you will see the recovery in Earth's ice first."

After the world emerges from the Covid pandemic that has killed millions of people, governments promise to 'build back better', to confront climate change, and the related crises we are causing to the natural world, from plastic pollution to deforestation and loss of species. The US decision to rejoin the Paris Agreement has given a glimpse of hope that the 2020s can be a time to transform the global economy, after decades of words. But time is fast running out, and there are alarming signs that an irreversible melt is under way at the ends of the Earth.

In Iceland, by the now-vanished Ok glacier on a mountaintop near Reykjavik, the "letter to the future" inscribed on a metal plaque warns that all the volcanic island's glaciers are likely to disappear in 200 years. Underscoring the planet-changing choices we face this decade, it says: "This monument is to acknowledge that we know what is happening, and what needs to be done. Only you know if we did it."

BIBLIOGRAPHY

Key sources are listed in chapter notes. The following are suggestions for further reading:

Björnsson, H., and Meldon D'Arcy, J. (translator), *The Glaciers of Iceland: A Historical, Cultural and Scientific Overview* (Amsterdam: Atlantis Press, 2016)

Ekman, M., *The Changing Level of the Baltic Sea During 300 Years: A Clue to Understanding the Earth* (Aaland Islands: Summer Institute for Historical Geophysics, 2009)

Englander, J., *Moving to Higher Ground* (Science Bookshelf, 2021)

Figueres, C., and Rivett-Carnac, T., *The Future We Choose* (New York: Penguin Random House, 2020)

Goodell, J., *The Water Will Come: Rising Seas, Sinking Cities and the Remaking of the Civilized World* (New York: Little, Brown and Co., 2017)

Kolbert, E., *Under a White Sky: The Nature of the Future* (New York: Crown, 2021)

McAdam, J., *Climate Change, Forced Migration, and International Law* (Oxford: Oxford University Press, 2012)

Masson-Delmotte, V., Zhai, P., Pirani, A., Connors, S.L., Péan, C., Berger, S., Caud, N., Chen, Y., Goldfarb, L., Gomis, M.I., Huang, M., Leitzell, K., Lonnoy, E., Matthews, J.B.R., Maycock, T.K.,

Waterfield, T., Yelekçi, O., Yu, R., and Zhou, B. (eds), IPCC, *Climate Change 2021: The Physical Science Basis – Contribution of Working Group I to the Sixth Assessment Report of the Intergovernmental Panel on Climate Change* (Cambridge, UK: Cambridge University Press, 2021)

Pilkey, O.H., and Pilkey, K.C., *Sea Level Rise: A Slow Tsunami on America's Shores* (North Carolina: Duke University Press Books, 2019)

Pörtner, H-O., Roberts, D.C, Masson-Delmotte, V., Zhai, P., Tignor, M., Poloczanska, E, Mintenbeck, K., Alegría, A., Nicolai, M., Okem, A., Petzold, J., Rama, B., and Weyer, N.M. (eds), IPCC, *2019: IPCC Special Report on the Ocean and Cryosphere in a Changing Climate* (New York: United Nations, 2019)

Rockstrom, J., and Gaffney, O., *Breaking Boundaries* (New York: DK, Penguin Random House, 2021)

Thorarinsson, T., *Present Glacier Shrinkage and Eustatic Changes of Sea-Level, Geografiska Annaler*, 1940, vol. 22:3–4, pp.131–59

Thunberg, G., *No One is Too Small to Make a Difference* (New York: Penguin Random House, 2019)

Wallace-Wells, D., *The Uninhabitable Earth* (New York: Tim Duggan Books, 2017)

ACKNOWLEDGEMENTS

THIS BOOK WAS ONLY POSSIBLE because of the kindness of strangers. I've been overwhelmed by the generosity of everyone I've met while writing this book – especially those on the frontlines in places from Fiji to Panama who have shared their experiences, anxieties and hopes for the future.

My thanks to everyone quoted: it's been uplifting to get so many perspectives – from people living by the shore, glaciologists, climate physicists, youth activists, politicians, engineers, lawyers, mountain guides, human rights experts, artists, poets and many others. I wish I'd been able to visit more places, especially in Asia and Africa, but the pandemic got in the way.

The Great Melt would never have reached you without the enthusiasm and skill of my editor Jo de Vries, at Flint Books, and the wonderful team: Alex Waite, Senior Project Editor; Katie Beard, Head of Design; Jemma Cox, Designer; Graham Robson, Campaigns Manager; and Cynthia Hamilton, Head of Marketing and PR.

Clare Grist Taylor, my agent at The Accidental Agency, tirelessly shaped what was a poorly focused idea about the swelling oceans into a coherent narrative. And my friend, author Roger Morgan-Grenville, was also key – he put me in touch with Clare.

I'm honoured that Christiana Figueres wrote the foreword; her blend of outrage and optimism that helped lead to the landmark

Paris Agreement on climate change in 2015 has been an inspiration for many years.

In Panama, thanks especially to Blas López who hosted me in his home on the Caribbean island of Gardi Sugdub. On Rabi Island in Fiji, David Christopher, Rosite Iotua and the Rabi Council were welcoming, letting me attend a memorable party that lasted almost until dawn. Simione Botu, thanks for letting me visit your home in Fiji, and for showing me what remains of the other two. In Peru, Saúl Luciano Lliuya kindly agreed to meet in a bar one busy day, and Noah Walker-Crawford gave valuable input on his legal case. In Iceland, thanks to Helgi Björnsson and Steinunn Sigurdardóttir for insightful feedback. Martin Ekman in Sweden has been an invaluable source for years about the history of the rising land around the Baltic Sea. Sjoerd Groeskamp, thanks for spending time reviewing the chapter about mega-dams to protect Europe.

I'm also grateful to my former employer, Thomson Reuters, for letting me write about the environment and climate change for many years, and to my colleague Stuart McDill who spotted that we could apply to go to Antarctica with the British Antarctic Survey. There we spent more than two weeks with some of the world's best climate scientists – thanks especially to David Vaughan and to Athena Dinar who became great guides and friends for us, far from home.

Also to Andrew Shepherd who helped put the planet's loss of ice into terms of personal ice cubes and Minecraft video games. And Anders Levermann, whose gripping way with words helped explain the thaw.

Behind the scenes, my thanks to many who've helped, especially Ed King and Ben Simonds.

My sister Margaret made many suggestions for improvements after reading drafts. Most of all, my gratitude and love to my wife, Siv, our daughter Emma and our son Matias, who have put up with me for years talking about sea level rise. Best of all were their reality checks: "Yawn" or "Really? Tell me more."

ENDNOTES

JOURNEYS INTO THE FRONTLINES OF CLIMATE CHANGE

1 UN Intergovernmental Panel on Climate Change (IPCC), *Climate Change 2021: The Physical Science Basis* (2021)
2 UN statement, 'Secretary-General Calls Latest IPCC Climate Report "Code Red for Humanity"', 9 August 2021
3 S.J. Arenstam Gibbons and R.J. Nicholls, 'Island abandonment and sea level rise: an historical analog from the Chesapeake Bay, USA', *Global Environmental Change* (2006)
4 Ocean: Find Your Blue, 'Sea level rise' (Smithsonian web resource)
5 International Energy Agency (IEA), press release: 'After a steep drop in early 2020, global carbon emissions have rebounded strongly' (2 March 2020)
6 World Meteorological Organization (WMO), '2020 was one of the warmest years on record' (15 January 2021)
7 World Meteorological Organization (WMO), '2020 was one of the warmest years on record' (15 January 2021)
5 www.worldweatherattribution.org/

CHAPTER 1

1 UN Intergovernmental Panel on Climate Change (IPCC), 'Special report – chapter 4: sea level rise and implications for low-lying islands, coasts and communities' (2019)
2 Thwaitesglacier.org, 'Thwaites glacier facts' (web resource)
3 UN Intergovernmental Panel on Climate Change (IPCC), *Climate Change 2021: The Physical Science Basis* (2021)

4 Ibid.
5 Ibid.
6 IPCC, 'Impacts of 1.5°C of global warming on natural and human systems' (2019)
7 National Science Foundation, 'Ice sheets' (web resource)
8 Testimony of Dr Richard B. Alley before the Committee of Science, Space and Technology – US House of Representatives, 'Antarctic contributions to future sea-level rise: possible tipping points' (11 July 2019)
9 EurekAlert!, news release: 'Antarctic ice shelves vulnerable to sudden meltwater-driven fracturing, says study' (26 August 2020)
10 AntarcticGlaciers.org, 'What is an ice shelf?' (web resource)
11 National Snow & Ice Data Center (NSIDC), news release: 'Ice bridge supporting Wilkins ice shelf collapses' (8 April 2009)
12 The European Space Agency, 'Wilkins ice shelf hanging by its last thread' (10 July 2008)
13 Britannica.com, 'History of Antarctica' (web resource)
14 United States Geological Survey, 'Paper 1386-B: satellite image atlas of glaciers of the world: Antarctica' (web resource)
15 NBC News, 'U.S.: "No time to lose" in Arctic, Antarctica' (6 April 2009)
16 M. Ranki and J.J. Fürst, 'Dynamic changes on the Wilkins Ice Shelf during the 2006–2009 retreat derived from satellite observations', *The Cryosphere* 11(3):1199–1211 (May 2017)
17 NASA Earth Observatory, 'Decline of West Antarctic Glaciers appears irreversible' (16 May 2014)
18 Eric Rignot declaration
19 Thwaitesglacier.org (web resource)
20 Ibid.
21 Thwaitesglacier.org, 'Thwaites glacier facts' (web resource)
22 *Science Daily*, news release: 'Antarctic ice shelves vulnerable to sudden meltwater-driven fracturing, says study' (26 August 2020)
23 Ibid.
24 Thwaitesglacier.org (web resource)
25 Ibid.
26 EurekAlert!, news release: 'Scientists find record warm water in Antarctica, pointing to cause behind troubling glacier melt' (29 January 2020)
27 Thwaitesglacier.org (web resource)
28 Thwaitesglacieroffshoreresearch.org, 'Blogs' (web resource)
29 British Antarctic Survey: Nunatak videos (YouTube resource)
30 WMO, 'New record for Antarctic continent reported' (14 February 2020)
31 IPCC, 'Fourth assessment report' (2007)
32 Reuters, 'Trump's view that ice caps "setting records" baffles scientists' (29 January 2018)
33 National Geographic, 'Ice cap' (web resource)
34 NSIDC: Scientific Data for Research (web resource)

CHAPTER 2

1 Secretariat of the Pacific Regional Environment Programme, news release: 'Upheaval of Fiji communities as climate change impacts are felt' (4 December 2014)
2 Fiji Low Emission Development Strategy 2018–2050
3 Ibid., p.5
4 Fiji's updated Nationally Determined Contribution
5 *New Zealand Herald*, 'Fiji village relocation is prime example to Bonn summit' (6 November 2017)
6 F. Gemenne, D. Ionesco and C. Zickgraf, *The state of environmental migration 2015: a review of 2014* (2015)
7 FBC News, 'Prince Harry reveals Vunidogoloa first village in the world to relocate' (24 October 2018)
8 Fiji's updated Nationally Determined Contribution
9 The official travel site of the Fiji Islands: www.fiji.travel/en
10 Fiji's updated Nationally Determined Contribution, p.9
11 The World Bank, 'Climate vulnerability assessment: making Fiji climate resilient' (1 October 2017)
12 United Nations Digital Library, 'Visit to Fiji: report of the Special Rapporteur' (2019)
13 Environmental Performance Index (EPI), '2020 EPI results' (web resource)
14 Clothilde Tronquet, *From Vunidogoloa to Kenani: an insight into successful relocation* (2015)
15 IslandsBusiness.com, news release: 'Vunidogoloa Rejects PM' (16 November 2018)
16 *Fiji Times*, 'Villagers branded as "liumuri" after no one voted for party' (11 November 2018)
17 Jane McAdam, 'Lessons from planned relocation and resettlement in the past', *Forced Migration Review Online* (May 2015)
18 A. Piggott-McKellar, K. McNamara, P. Nunn and S. Sekinini, 'Moving people in a changing climate: lessons from two case studies in Fiji', *Social Sciences* (2019)
19 National Legislative Bodies / National Authorities, 'Fiji: Planned Relocation Guidelines: a framework to undertake climate change related relocation' (2018)
20 IPCC, 'Special report – chapter 4: sea level rise and implications for low-lying islands, coasts and communities' (2019)
21 UN News, 'UN environmental body hails relocation of islanders threatened by climate change' (6 December 2005)
22 Equator Initiative / UNDP, 'Equator Initiative case studies: Tulele Peisa' (2016)
23 Sarah M. Munoz, 'Understanding the human side of climate change relocation', The Conversation (5 June 2019)
24 World Economic Forum, 'Five places relocating people because of climate change' (29 June 2017)

25 UNFCCC, Frank Bainimarama speech (18 May 2017)
26 Climate Vulnerable Forum
27 Fiji's Constitution
28 Reuters, 'As seas rise, Pacific island president favours buying land abroad' (22 September 2014)
29 *Science* magazine, 'Thresholds of mangrove survival under rapid sea level rise' (5 June 2020)
30 P.S. Kench, M.R. Ford and S.D. Owen, 'Patterns of island change and persistence offer alternate adaptation pathways for atoll nations', *Nature Communications*, 9, 605 (2018)
31 The Rush Limbaugh Show, 'More evidence climate change is fake' (9 February 2018)
32 *Fiji Times*, 'Tuvalu PM refutes AUT research' (12 February 2018)
33 L. Friedman, 'Tuvalu is growing (for now, at least)', *New York Times* (14 February 2018)
34 IPCC, 'Special report' (2019)
35 Antigua and Barbuda News, 'Address by Prime Minister Gaston Browne at UN Security Council Open Debate on climate change' (23 February 2021)

CHAPTER 3

1 The SAO/NASA Astrophysics Data System, 'Revisiting the catastrophic 1941 outburst flood of lake Palcacocha' (2018), 20th EGU General Assembly, EGU2018, Proceedings from the conference held 4–13 April 2018 in Vienna, Austria, p.7569
2 NASA Earth Observatory, 'Glacial collapse threatens Huaraz, Peru' (5 November 2001)
3 IPCC, 'Special report – chapter 2: high mountain areas' (2019)
4 Germanwatch.org, 'Overview of the case of Huaraz' (web resource)
5 Ibid.
6 Ibid.
7 Somos-Valenzuela, Marcelo A., *et al.*, 'Inundation modeling of a potential Glacial Lake Outburst Flood', Center for Research in Water Resources, University of Texas at Austin (2014)
8 Germanwatch.org, 'Court documents of the "Huaraz Case"' (web resource)
9 M. Mergili, *et al.*, 'Reconstruction of the 1941 GLOF process chain at Lake Palcacocha (Cordillera Blanca, Peru)', *Hydrology and Earth System Sciences* (2020)
10 National Meteorology and Hydrology Service, 'Climate Scenarios for Peru to 2030'
11 IPCC, 'Special report – chapter 2: high mountain areas' (2019)
12 Germanwatch.org, 'Translation of Higher Regional Court of Hamm'
13 Environmental Law Alliance Worldwide, 'Native Village of Kivalina v. ExxonMobil Corp.' (2012)
14 Germanwatch.org, 'Translation of District Court of Essen'

15 R.F. Stuart-Smith, *et al.*, 'Increased outburst flood hazard from Lake Palcacocha due to human-induced glacier retreat', *Nat. Geosci.* 14, 85–90 (2021)
16 ScienceMag.org, 'A massive rock and ice avalanche caused the 2021 disaster at Chamoli, Indian Himalaya' (16 July 2021)
17 P. Stott, *et al.*, 'Human contribution to the European heatwave of 2003', *Nature* 432, 610–614 (2004)
18 J.-M. Robine, *et al.*, 'Death toll exceeded 70,000 in Europe during the summer of 2003', *C.R. Biologies* 331 (2008)
19 Carbon Brief, 'Mapped: How climate change affects extreme weather around the world' (web resource)
20 Global Monitoring Laboratory, 'Trends in Atmospheric Carbon Dioxide' (web resource)
21 UNEP, press release: 'Surge in court cases over climate change shows increasing role of litigation in addressing the climate crisis' (26 January 2021)
22 Urgenda, 'Landmark decision by Dutch Supreme Court' (20 September 2019)
23 Our Children's Trust, 'Overview of the Florida case'
24 Court complaint, Leon County, Florida: *Delaney Reynolds vs. The State of Florida*
25 United States Nationally Determined Contribution
26 Order Granting Motions to Dismiss with Prejudice: *Delaney Reynolds vs. The State of Florida*
27 Our Children's Trust, '350ppm Pathways report' (2019)
28 US Supreme Court docket, United States, *et al.*, *Applicants v. United States District Court for the District of Oregon* (17 July 2018)
29 Vatican, Pope Francis speech 2019
30 French Ministry of Ecological Transition, 'Loi climat et résilience: l'écologie dans nos vies' (14 June 2021)
31 Ruling by the First Senate of the Federal Constitutional Court
32 Ruling by the Hague District Court
33 Freshfields Bruckhaus Deringer, 'A new front in the fight against climate change' (web resource)
34 Germanwatch.org, 'Overview of the case of Huaraz' (web resource)
35 Climate Accountability Institute, press release: 'Update on Carbon Majors 1965–2018' (9 December 2020)
36 Public Papers of the Presidents of the United States: Lyndon B. Johnson (1965)
37 US government, 'Restoring the quality of our environment: report of the Environment Pollution Panel, the President's Scientific Advisory Committee' (5 November 1965)
38 UNFCCC, Paris Agreement text
39 IPCC, 'Special report: summary for policymakers' (2019)
41 NASA Earth Observatory, 'Glacial collapse threatens Huaraz, Peru' (5 November 2001)

CHAPTER 4

1 IPCC, 'Special report on the ocean and cryosphere in a changing climate' (2019)
2 France24, 'Oceans turning from friend to foe, warns landmark UN climate report' (29 August 2019)
3 UN Intergovernmental Panel on Climate Change (IPCC), *Climate Change 2021: The Physical Science Basis* (2021)
4 Malé Declaration
5 The Nobel Peace Prize 2007
6 Yale Climate Connections, 'Anatomy of IPCC's mistake on Himalayan glaciers and year 2035' (4 February 2010)
7 *Earth Negotiations Bulletin*, 'Highlights and images of main proceedings for 20 September 2019' (20 September 2019)
8 Saudi Press Agency, 'HRH Crown Prince announces: "The Saudi Green Initiative and The Middle East Green Initiative"' (27 March 2021)
9 Carbon Brief, 'The Carbon Brief Interview: Saudi Arabia's Ayman Shasly' (12 December 2018)
10 S. Schneider, *Science as a contact sport: inside the battle to save the Earth's climate* (National Geographic, 2009)
11 Climate One, 'My Climate Story: Ben Santer' (17 September 2019)
12 IPCC, press release: 'Choices made now are critical for the future of our ocean and cryosphere' (25 September 2019)
13 IPCC, 'Special report – chapter 4: sea level rise and implications for low-lying islands, coasts and communities' (2019), p.336
14 Yacht Harbour, 'Top league charter: 111m superyacht *TIS* for EUR 2.2 million a week' (26 July 2019)
15 UNFCCC, Monaco's greenhouse gas emissions
16 enb.iisd.org/climate/IPCC/IPCC-54-WGI-14/summary?utm_medium=email&utm_campaign=ENB%20Update%20-%209%20August%202021&utm_content=ENB%20Update%20-%209%20August%202021+CID_9660b06859baedb9229a300b1110386f&utm_source=cm&utm_term=Read#brief-analysis-ipcc-54

CHAPTER 5

1 Jane McAdam, 'Caught between homelands', insidestory.org.au
2 Julia B. Edwards, 'Phosphate mining and the relocation of the Banabans to northern Fiji in 1945: lessons for climate change-forced displacement', *Journal de la Société des Océanistes* (2014)
3 *The Fiji Times*, 'Finding a new home away from home' (30 December 2015)
4 UK Parliament Hansard Debate, 'Kiribati Bill Hl' (19 February 1979)
5 Jane McAdam / UNFCCC, '"Under Two Jurisdictions": immigration, citizenship, and self-governance in cross-border community relocations' (n.d.)
6 Kiribati parliamentary records (20 August 2020)

7 TE ARA – The Encyclopaedia of New Zealand, 'Biography of Albert Ellis' (web resource)
8 Jane McAdam, 'Caught between homelands', insidestory.org.au
9 BBC, 'Go tell it to the judge' (YouTube resource)
10 Ibid.
11 UK Parliament Hansard Debate, 'Banaban compensation claims', (26 July 1979)
12 Katerina Martina Teaiwa, *Consuming Ocean Island: stories of people and phosphate from Banaba* (2015)
13 Asia-Pacific Network for Global Change, 'Community relocation as an option for adaptation to the effects of climate change and climate variability in Pacific Island Countries (PICs)' (2005)
14 'Memorandum on the future of the Banaban population' (2 September 1946)
15 UK Parliament Hansard Debate, 'Ocean Island and Banabans' (27 May 1977)
16 UNFCCC, 'Fourteenth session of the Conference of Parties (COP14)' (11 December 2008)
17 Reuters, 'As seas rise, Pacific island president favours buying land abroad' (22 September 2014)
18 Asia-Pacific Network for Global Change, 'Community relocation as an option for adaptation to the effects of climate change and climate variability in Pacific Island Countries (PICs)' (2005)
19 'Agreement between the United States and Cuba for the lease of lands for coaling and naval stations; February 23, 1903' (Yale Law School, Lilian Goldman Law Library collection)
20 G.M. Tabucanon and B. Opeskin, 'The resettlement of Nauruans in Australia: an early case of failed environmental migration', *The Journal of Pacific History*, vol. 46, no. 3 (December 2011)

CHAPTER 6

1 Displacement Solutions, 'One step at a time: the relocation process of the Gardi Sugdub community in Gunayala, Panama' (2015)
2 UN Intergovernmental Panel on Climate Change (IPCC), *Climate Change 2021: The Physical Science Basis* (2021)
3 N. Cortizo, Facebook presidential video (in Spanish) (11 September 2020) (web resource)
4 República de Panamá, press release: 'Ministry of Housing resumes construction of Nuevo Carti in Guna Yala (statement in Spanish)' (19 October 2020)
5 Reuters, 'Panama's new president takes office, pledges end to corruption' (1 July 2019)
6 President Cortizo, Twitter statement (in Spanish) (15 September 2020)
7 BBC, 'The Darien Venture' (web resource)
8 John Prebble, *Darien: The Scottish Dream of Empire* (Birlinn, 2020)

9 República de Panamá, 'Panamanian National Assembly law (in Spanish)' (2011)
10 Inter-American Development Bank, 'Sustainable movement of Guna due to climate change (in Spanish)' (November 2018)
11 gunayala.com, 'The history of Guna Yala and its people' (web resource)
12 M.M. Mauri, 'Kuna Yala, tierra de mar (in Spanish)' (2011)
13 IPCC, 'Summary for policymakers of IPCC Special Report on global warming of 1.5 degrees C approved by governments' (2018)
14 H.M. Guzmán, *et al.*, 'Natural Disturbances and Mining of Panamanian Coral Reefs by Indigenous People', *Conservation Biology* (2003)
15 Inter-American Development Bank, 'Sustainable movement of Guna due to climate change' (in Spanish) (November 2018)
16 Displacement Solutions, 'The Peninsula Principles in action: Panama mission report' (2014)
17 UNFCCC, 'Panama's updated Nationally Determined Contribution' (in Spanish)
18 Inter-American Development Bank, 'Document outlining school plan' (in Spanish) (n.d.)
19 UN Panamá, 'National Climate Change Strategy for 2050' (in Spanish) (2020)
20 telemetro.com, 'En el olvido se encuentra' (12 September 2020)
21 The World Bank / ICSID, 'Omega Engineering LLC and Oscar Rivera v. Republic of Panama' (updated 21 January 2021)

CHAPTER 7

1 Martin Ekman, 'An investigation of Celsius' pioneering determination of the Fennoscandian Land Uplift Rate, and of his Mean Sea Level Mark', Summer Institute for Historical Geophysics Åland Islands www.historicalgeophysics.ax (2013)
2 Martin Ekman, 'The changing level of the Baltic Sea during 300 years: a clue to understanding the Earth', Summer Institute for Historical Geophysics Åland Islands www.historicalgeophysics.ax (2009)
3 Martin Ekman, 'The man behind "Degrees Celsius": a pioneer in investigating the Earth and its changes', Summer Institute for Historical Geophysics Åland Islands www.historicalgeophysics.ax (2016)
4 John Playfair, 'Illustrations of the Huttonian theory of the Earth' (Edinburgh, 1802) – can be sourced at: www.biodiversitylibrary.org
5 Charles Lyell, 'The Bakerian Lecture: on the proofs of a gradual rising of the land certain parts of Sweden' (1 January 1835)
6 Ibid.
7 Ibid.
8 Ibid.
9 Ibid.
10 Ibid.
11 'People of Science with Brian Cox: Richard Fortey on Charles Lyell' (5 January 2020) (YouTube resource)

12 Charles Lyell, *Principles of Geology* (London, 1830–33)
13 The Lyell Centre, 'Sir Charles Lyell' (web resource)
14 illinois.edu, 'The life and work of Louis Agassiz' (web resource)
15 The Smithsonian Institution, 'Through time: ancient seas, sea-level rise' (web resource)
16 Thomas F. Jamieson, 'On the history of the last geological changes in Scotland', *Quarterly Journal of the Geological Society* (1 February 1865)
17 UN Intergovernmental Panel on Climate Change (IPCC), *Climate Change 2021: The Physical Science Basis* (2021)
18 Britannica.com, 'Superstorm Sandy' (web resource)
19 Henrik Vuorinen / Port of Luleå, 'The sustainable link to the world' (2019)
20 coastalratepayersunited.co.nz, 'Sea level is not rising' (n.d.)
21 Ibid.
22 George Monbiot's blog, 'The Spectator runs false sea-level claims on its cover', *The Guardian* (2 December 2011)
23 *Daily Mail*, 'Doomsday predictions on sea level rises are "false alarm" – levels always fluctuate, says expert as climate change row heats up' (1 December 2011)
24 INQUA Neotectonics Commission, INQUA Shorelines Commission, the IGCP 437 Sea Level Project and the ILP II-5, 'The Sweden excursion' (May 1999)

CHAPTER 8

1 Icelandic Meteorological Office, *et al.*, 'Overview of Icelandic Glaciers at the End of 2019' (2020)
2 NASA Earth Observatory, 'Glossary' (web resource)
3 *Frontiers in Earth Science*, 'Glacier changes in Iceland from 1890 to 2019' (26 November 2020)
4 US Department of the Interior / US Geological Survey, 'Geographic names of Iceland's glaciers: historic and modern' (2008)
5 Ari Þorgilsson and Thomas Ellwood, *The Book of the Settlement of Iceland* (1898)
6 Jonathan Grove, *The Place of Greenland in Medieval Icelandic Saga Narrative* (2009)
7 John Nichol, *Byron* (eBook)
8 Andri Snær Magnason, 'The glaciers of Iceland seemed eternal. Now a country mourns their loss', *The Guardian* (14 August 2019)
9 Not OK website: www.notokmovie.com/
10 *The Economist*, 'Obituary: Okjökull was declared dead in 2014' (21 September 2019)
11 Martin Stendel tweet (2 August 2019)
12 Katrín Jakobsdóttir, 'Iceland's Prime Minister: "The Ice is Leaving"', *The New York Times* (17 August 2019)
13 Helgi Björnsson, 'Glaciers in Iceland' (Google Scholar citations)

14 UN, 'Bolivia's glacier: a vanishing future' (2009)
15 BBC, 'Adiós al glaciar Chacaltaya' (6 August 2009)
16 World Glacier Monitoring Service (WGMS), 'Global glacier change bulletin' (2016–17)
17 Ibid.
18 Climate-ADAPT, 'Hydropower expansion and improved management in response to increased glacier melt in Iceland' (n.d.)
19 Landsvirkjun, 'Annual report' (2018)
20 Helgi Björnsson, *The Glaciers of Iceland* (2017)
21 Avijit Gupta, *Large Rivers: Geomorphology and Management* (2007)
22 Pen America, 'Island on Fire: The Extraordinary Story of a Forgotten Volcano that Changed the World' pen.org (25 March 2016)
23 Carolina Pagli and Freysteinn Sigmundsson, 'Will present day glacier retreat increase volcanic activity? Stress induced by recent glacier retreat and its effect on magmatism at the Vatnajökull ice cap, Iceland', *Solid Earth* (7 May 2008)
24 Secretariat of the Convention on Biological Diversity, 'Geoengineering in relation to the Convention on Biological Diversity' (September 2012)
25 An Interview with Professor Sigurdur Thorarinsson (YouTube resource) (14 April 2016)
26 Sigurdur Thorarinsson, 'Present glacier shrinkage, and eustatic changes of sea-level', *Geografiska Annaler* (29 August 2017)
27 Helgi Björnsson, 'Sigurdur Thorarinsson', *Journal of Glaciology* (20 January 2017)

CHAPTER 9

1 Sjoerd Groeskamp and Joakim Kjellsson, 'NEED: The Northern European Enclosure Dam for if climate change mitigation fails', *Bulletin of the American Meteorological Society* (3 August 2020)
2 UN Intergovernmental Panel on Climate Change (IPCC), *Climate Change 2021: The Physical Science Basis* (2021)
3 US Army Corps of Engineers, 'West Closure Complex' (August 2015)
4 Rory Clisby and Will Nichols, 'Environmental Risk Outlook 2020', Verisk Maplecroft (27 February 2020)
5 The Sand Motor: dezandmotor.nl/en/ (web resource)
6 Jacqueline Heerema, 'Timely reflections' (web resource)
7 North Norfolk Council, 'Bacton to Walcott sandscaping'
8 North Norfolk Council, 'Sandscaping: frequently asked questions'
9 Matto Mildenberger, Jennifer Marlon, Peter Howe and Anthony Leiserowitz, *Democratic and Republican Views of Climate Change* (2018)
10 *The New York Times*, 'The $119 billion sea wall that could defend New York ... or not' (17 January 2020)
11 Anna Weber, 'The Army Corps can't save Miami from climate change', Natural Resources Defense Council, www.nrdc.org (24 August 2020)

12 Miami Dade County, 'Sea Level Rise Strategy'
13 A.R. Siders and Jesse M. Keenan, 'Variables shaping coastal adaptation decisions to armor, nourish, and retreat in North Carolina', *Ocean & Coastal Management* (1 January 2020)
14 Beach Nourishment Viewer: beachnourishment.wcu.edu/ (web resource)
15 National Bureau of Economic Research, 'Neglected no more: housing markets, mortgage lending, and sea level rise' (October 2020)
16 Union of Concerned Scientists, 'Underwater: rising seas, chronic floods, and the implications for US coastal real estate' (18 June 2018)
17 European Geosciences Union, 'Delaying future sea-level rise by storing water in Antarctica', *Earth System Dynamics* (2016)
18 WMO, press release: '2020 was one of the three warmest years on record' (15 January 2021)
19 Secretariat of the Antarctic Treaty, 'The Protocol on Environmental Protection to the Antarctic Treaty'
20 John C. Moore, *et al.*, 'Geoengineer polar glaciers to slow sea-level rise', *Nature* (14 March 2018)
21 The National Academies of Science, Engineering and Medicine, 'Reflecting sunlight: recommendations for Solar Geoengineering research and research governance' (March 2021)

CHAPTER 10

1 UN, 'International Covenant on Civil and Political Rights'
2 Caroline Sawyer, 'The loss of birthright citizenship in New Zealand' (n.d.)
3 *Ioane Teitiota v. The Chief Executive of the Ministry of Business, Innovation and Employment*, New Zealand Supreme Court ruling (20 July 2015)
4 COP23.com, 'Kiribati and a changing climate'
5 UN Human Rights: Office of the High Commissioner, press release: 'Historic UN Human Rights case opens door to climate change asylum claims' (21 January 2020)
6 Reuters, 'World needs to prepare for "millions" of climate displaced: U.N.' (21 January 2020)
7 Evan Wusaka, 'Landmark decision from UN Human Rights Committee paves way for climate refugees', ABC News (21 January 2020)
8 Raoul Wallenberg Institute of Human Rights and Humanitarian Law in Lund, Sweden
9 'Rising Sea Levels: Promoting Climate Justice through International Law: British Institute of International and Comparative Law' (YouTube resource) (18 March 2021)
10 WMO, 'State of the Global Climate 2020' (2021)
11 International Federation of Red Cross and Red Crescent Societies (IFRC), 'The cost of doing nothing' (2019)

12 Colin P. Kelley, *et al.*, 'Climate change in the fertile crescent and implications of the recent Syrian drought', *Proceedings of the National Academy of Sciences of the United States of America* (17 March 2015)
13 UNHCR, 'Climate change and disaster displacement' (23 December 2019)
14 IPCC, 'How many people could be displaced as a result of climate change?' (2018)
15 UNHCR, 'The 1951 Refugee Convention'
16 Judgement of Priestley, J., High Court of New Zealand, Auckland Registry
17 NBC News, 'Cabinet makes splash with underwater meeting' (17 October 2009)
18 UN, 'United Nations Convention on the Law of the Sea'
19 The Pacific Community (web resource)
20 UNDP, Pacific Office in Fiji (web resource)
21 Western and Central Pacific Fisheries Commission (web resource)
22 Kevin Rudd, 'The complacent country' (4 February 2019)
23 ABC News, 'Tuvalu PM slams Kevin Rudd's proposal' (17 February 2019)
24 Bruce Hill, 'Sovereignty for citizenship might help the Pacific', The Lowy Institute (28 February 2019)
25 International Law Association
26 International Law Association, 'Resolution 5' (2018)
27 Pacific Islands Forum, 'Simon Kofe speech' (9 September 2020)
28 Permanent Court of Arbitration, 'The South China Sea Arbitration' (web resource)
29 Ibid.
30 Clive Schofield and David Freestone, 'Islands awash amidst rising seas: sea level rise and insular status under the Law of the Sea', *The International Journal of Marine and Coastal Law* (30 August 2019)
31 UN, 'Montevideo Convention'
32 'Rising sea levels: promoting climate justice through international law webinar' (YouTube resource) (4 March 2021)
33 R. Rayfuse and S. Scott, 'Climate change, sovereignty and statehood' (18 November 2014)
34 Order of Malta (web resource)
35 BBC, 'The man who would be the first climate change refugee' (5 November 2015)

EPILOGUE

1 European Geosciences Union, 'Review: Earth's ice imbalance', *The Cryosphere* (25 January 2021)
2 The European Space Agency, 'In perspective: 1 trillion tonnes of ice' (web resource)
3 World Resources Institute, '4 charts explain greenhouse gas emissions: countries and sectors' (web resource)
4 F.R. Rijsberman and R.J. Swart, 'Targets and indicators of climate change', The Stockholm Environment Institute (1990)

INDEX